Editorial Policy
for the publication of monographs

In what follows all references to monographs, are applicable also to multiauthorship volumes such as seminar notes.

§1. Lecture Notes aim to report new developments - quickly, informally, and at a high level. Monograph manuscripts should be reasonably self-contained and rounded off. Thus they may, and often will, present not only results of the author but also related work by other people. Furthermore, the manuscripts should provide sufficient motivation, examples and applications. This clearly distinguishes Lecture Notes manuscripts from journal articles which normally are very concise. Articles intended for a journal but too long to be accepted by most journals, usually do not have this "lecture notes" character. For similar reasons it is unusual for Ph. D. theses to be accepted for the Lecture Notes series.

§2. Manuscripts or plans for Lecture Notes volumes should be submitted (preferably in duplicate) either to the editor of the series or to Springer-Verlag, Heidelberg. These proposals are then refereed. A final decision concerning publication can only be made on the basis of the complete manuscript, but a preliminary decision can often be based on partial information: a fairly detailed outline describing the planned contents of each chapter, and an indication of the estimated length, a bibliography, and one or two sample chapters - or a first draft of the manuscript. The editor will try to make the preliminary decision as definite as he can on the basis of the available information.

§3. Final manuscripts should be in English. They should contain at least 100 pages of scientific text and should include
- a table of contents;
- an informative introduction, perhaps with some historical remarks: it should be accessible to a reader not particularly familiar with the topic treated;
- a subject index: as a rule this is genuinely helpful for the reader.

Further remarks and relevant addresses at the back of this book.

Lecture Notes in Biomathematics   94

Ross Cressman

The Stability Concept
of Evolutionary Game Theory

A Dynamic Approach

Springer-Verlag

Berlin Heidelberg New York
London Paris Tokyo
Hong Kong Barcelona
Budapest

Author

Ross Cressman
Department of Mathematics
Wilfrid Laurier University
Waterloo, Ontario, Canada N2L 3C5

Mathematics Subject Classification (1980): 92DXX, 90DXX, 34DXX

ISBN 978-3-540-55419-6   ISBN 978-3-642-49981-4 (eBook)
DOI 10.1007/978-3-642-49981-4

© Springer-Verlag Berlin Heidelberg 1992
Typesetting: Camera ready by author/editor
46/3140-543210 - Printed on acid-free paper

Preface

These Notes grew from my research in evolutionary biology, specifically on the theory of evolutionarily stable strategies (ESS theory), over the past ten years. Personally, evolutionary game theory has given me the opportunity to transfer my enthusiasm for abstract mathematics to more practical pursuits. I was fortunate to have entered this field in its infancy when many biologists recognized its potential but were not prepared to grant it general acceptance. This is no longer the case. ESS theory is now a rapidly expanding (in both applied and theoretical directions) force that no evolutionary biologist can afford to ignore.

Perhaps, to continue the life-cycle metaphor, ESS theory is now in its late adolescence and displays much of the optimism and exuberance of this exciting age. There are dangers in writing a text about a theory at this stage of development. A comprehensive treatment would involve too many loose ends for the reader to appreciate the central message. On the other hand, the current central message may soon become obsolete as the theory matures. Although the restricted topics I have chosen for this text reflect my own research bias, I am confident they will remain the theoretical basis of ESS theory. Indeed, I feel the adult maturity of ESS theory is close at hand and I hope the text will play an important role in this achievement.

The text will be of interest to theoretical and applied evolutionary biologists, applied mathematicians and game theorists. The different languages of these disciplines mean most readers will be more comfortable with some sections than with others. In particular, the mathematical content varies greatly from section to section. Sections 2.8 to 2.10 can be omitted on first reading by those who prefer to avoid mathematical technicalities as long as possible.

It is a pleasure to thank colleagues who have helped my own growth in evolutionary game theory; in particular, Ethan Akin, Tim Bishop, Art Dash, John Haigh, Peter Hammerstein, Gord Hines, Sabin Lessard, Gordon Prior, Peter Taylor and Glenn Vickers. I am also indebted to Ethan Akin, Barb Carroll, Karen Cressman, Tina Keeble and Marc Kilgour for their comments on earlier versions of the text. Assistance by June Aleong, Pam Schaus and especially Jane Schmalz with the manuscript preparation is gratefully acknowledged. A special thanks is also extended to Simon Levin, my family and friends who all encouraged me in this project. Karen was particularly patient and my children kept it in proper perspective, assuming by the time I finished that I must be working on a second book. Financial assistance was received from Wilfrid Laurier University and from the Natural Sciences and Engineering Research Council of Canada.

Waterloo, Ontario
January, 1992

Contents

1. INTRODUCTION

Over the past two decades, evolutionary game theory has become a powerful unifying force to analyze evolutionary processes that are driven by individual selection pressures. It is now the accepted theoretical explanation of such diverse biological phenomena as the prevalence of equal male/female sex ratios in diploid species and the evolution of ritualistic fighting behaviors in many animal species (Maynard Smith 1982; Hofbauer and Sigmund 1988). A random check of current research publications in (theoretical) biology shows both the growing breadth and depth of evolutionary game theory's influence. American Naturalist in 1991 published articles that extend the theory of evolutionary games (Christiansen 1991); its range of biological applications (Motro 1991); and its consistency with empirical data (Petersen 1991). Even more convincing for me is that the majority of articles in these volumes contain some form of strategic reasoning (e.g. the analysis of individual behavioral alternatives in a given biological system).

In its early form (Maynard Smith and Price 1973; Maynard Smith 1974), single-species evolutionary game theory relied heavily on classical (i.e. non-biological) game theory. Here individuals make a strategic choice that is based on rational decisions concerning its payoff consequences. A crucial step for evolutionary applications was the realization that an individual's payoff can be equated to its reproductive success — thereby avoiding the questionable assumption that biological populations make rational decisions. In hindsight, it is clear that the static solution concepts from classical non-cooperative games lead naturally to the heuristic definition of an evolutionarily stable strategy (ESS). That is, an ESS is a strategy that cannot be successfully invaded by a rare mutant strategy (Maynard Smith 1982).

At the time, the mathematical proof that an ESS is noninvadable depended on a dynamic (Taylor and Jonker 1978) that assumed an individual produced offspring who were identical to their parent. In particular, any genetic effects due to mating in diploid populations were summarily ignored. Unfortunately, this simplification and others (e.g. ignoring the potential density effects due to a habitat's limited resources) confirmed, for many skeptics, that evolutionary game theory was little more than a heuristic tool that could at best suggest evolutionary tendencies but never prove them convincingly in bona fide biological models.

This text emphatically refutes such criticism by showing the noninvadability of an ESS in models that incorporate genetic and/or density effects. Moreover, I feel the trend of theoretical evolutionary games moving closer to biological reality will continue, though in what direction is unclear. Perhaps the greatest potential lies in multi-species coevolutionary models mentioned briefly in different parts of the text.

A great deal of the text is devoted to the development of static ESS conditions in diverse theoretical models of evolutionary biology. Indeed, an "ESS program" is introduced and implemented in all these models to emphasize their similarities. The program also includes the analysis of dynamic stability for ESS's in various biological contexts — in my view the main purpose of this text. The combined emphasis on both static and dynamic properties of theoretical evolutionary games distinguishes this text from other books on evolutionary game theory. Perhaps the only book with a more complete theoretical treatment of any of the models is Bomze and Pötscher (1989). Their treatment of single-species, frequency-dependent haploid evolution considers more general (non-matrix) evolutionary games where the frequencies of individual strategies may form a continuous distribution.

Throughout this text, there are a finite number of types of individuals (i.e. phenotypes) in the biological system and each phenotypic fitness is given by a random interaction with other individuals. Game theory enters each model by way of payoff matrices that quantify individual fitness. Examples of evolutionary games are chosen to illustrate the theory first and, to a lesser extent, to reflect real applications.

No attempt is made to justify ESS theory on empirical grounds or, indeed, to suggest ways one might carry out such a justification. Nonetheless, this text is not solely intended for the converted. It is hoped that the models considered here will be sufficient to convince the most skeptical reader, whether inclined towards theory or applications, that evolutionary game theory deserves a central chapter in the book of evolutionary biology.

Throughout, static ESS conditions are developed hand-in-hand with a thorough analysis of the underlying evolutionary dynamic. As implied above, this is contrary to the historical development of ESS theory that emphasized a more heuristic static approach to biological stability (Maynard Smith and Price 1973). Only later was dynamic stability included. The historical sequence is entirely appropriate for applied evolutionary game theory because the main benefit for applications is the principle that the evolutionary outcome of selection can be predicted by ESS theory without recourse to the complex mathematical structure on which most biologists feel these predictions ultimately rest. In contrast, this text is based on the premise that, for theoretical evolutionary games, the dynamic must be considered a priori to substantiate this principle. In fact, my personal opinion is that the main value of theoretical evolutionary games from a biological perspective is the justification of these heuristic shortcuts that are so often used in applications.

Chapters 2 and 4 both deal with frequency-dependent selection in a single species. The equivalence of ESS theory and dynamic stability is most clear in the haploid model of Chapter 2. On the other hand, the inclusion of diploid random mating in Chapter 4 already introduces theoretical difficulties. It is through a precise understanding of these difficulties that the power of evolutionary game theory can best be appreciated. Both chapters rely heavily on the existing literature. At the same time, it is hoped that the unrelenting insistence on a thorough dynamical analysis of the model at hand and the attendent development of the unifying structure (e.g. the strong stability concept) of ESS theory will provide sufficient interest for the expert in the field.

Chapters 3 and 5 present a novel approach to evolutionary stability for two-species interactions and for single-species density-dependent interactions respectively. These chapters reflect the view that stability in ecological models can be based on game-theoretic aspects of individual competition (inter and/or intraspecific). As such, they are part of the theory of coevolution (Roughgarden 1979) but here the emphasis is on the intuitive understanding of stability made possible by a careful analysis of the dynamic associated with the evolutionary game.

Each chapter from 2 to 5 can be summarized as the implementation of the ESS program (see Section 2.6) for a particular theoretical model in evolutionary biology. In fact, the order of these chapters is essentially arbitrary in that the independent analysis of any one of them leads to the same program.

Chapter 6 is different both in substance and in style. This last chapter adapts the previous theory to fit into new applications (e.g. the owner-intruder game and the iterated prisoner's dilemma game). The additional theory of evolutionarily stable sets (ES Sets) and extensive evolutionary games introduced here is carefully limited to what is needed to analyze the application at hand. This application-driven approach parallels the historical development of evolutionary game theory. Readers familiar with mathematical modelling (Mesterton-Gibbons 1992) will be comfortable with the approach. It is adopted here to emphasize that game-theoretic techniques can often give qualitative answers to questions in behavioral biology that seem intractable by conventional optimization and/or dynamic methods. I feel strongly that an appreciation of this important fact will strengthen the heuristic understanding of the previous theory.

This text should not be regarded as a review either of theoretical evolutionary games (for such a review, see Hines (1987)) or of evolutionary dynamics (Hofbauer and Sigmund 1988). Rather, its purpose is to examine carefully the power and limitations of ESS theory in a restricted but central portion of evolutionary biology. The sentiment contained in the afterword of Hofbauer and Sigmund (1988) is just as appropriate here; that is, omitted completely are all such practical biological considerations as mutation rates; dispersal and migration rates; spatial effects; effects of finite population size; assortative interactions; determination of quantitative behaviors and fitness parameters; stochastic effects etc...... But what can one do? That's life.

2. FREQUENCY-DEPENDENT EVOLUTION IN A SINGLE HAPLOID SPECIES

2.1 Pure and Mixed Strategies

Most researchers in the area agree that the approach of Maynard Smith (1974; 1982) to model the behavioral evolution in a single animal species was the beginning of evolutionary game theory. In this approach, individuals are characterized by their behavior type (or *strategy*). These strategies may be bona fide behaviors such as aggression or altruism that can be used in pairwise interactions; physical characteristics such as size or color; or life-cycle patterns such as dispersal rate or sex ratio. An individual with a strategy of the latter type is often thought of as playing the field in that payoffs are not given through pairwise contests. In any case, an individual's strategy is fixed over its lifetime or, alternatively, the life history of an individual is its strategy. [This assumption contrasts markedly with that of classical (i.e. economic or non-biological) game theory where individuals may change their strategies based on rational decisions concerning alternative payoffs.] An individual's strategy, in conjunction with the prevalence of strategies of other individuals, determines its frequency-dependent fitness and thereby its reproductive success. [For further elaboration on these points, see Hofbauer and Sigmund (1988, Section 15.3).]

All the models in this text are based on a finite number of behavior types (see Introduction). The *finite-strategy model* assumes that, at any particular instant, an individual must exhibit one of the m possible behaviors $\{\pi_1, \pi_2, \ldots, \pi_m\}$, say π_k. Moreover, in the *pure-strategy submodel*, such an individual exhibits the pure strategy π_k at all times. Individuals who use *mixed strategies* also appear throughout the text. For a single species, such an individual may use π_k some of the time and other pure strategies at other times but the proportion s_k of the time a particular individual uses π_k is fixed. That is, in a mixed-strategy model, an individual is characterized by a probability vector

$$S \in \Delta^m = \left\{ (s_1, s_2, \ldots, s_m) \,\Big|\, \sum_{k=1}^{m} s_k = 1, \ s_k \geq 0 \right\} \tag{2.1.1}$$

where s_k is the probability it uses the kth pure strategy at a given instant. S is called the individual's strategy or *phenotype*. The pure-strategy model is then a special case of the mixed-strategy model where the pure strategy π_k is now represented by the unit coordinate vector $e_k = (0, \ldots, 0, 1, 0, \ldots 0)$ that has 1 in the kth component.

In either model, the number of different phenotypes is assumed to be finite — thus ruling out a continuum of pure strategies such as one might wish in a behavior based on size or dispersal rate. This is not as severe a restriction as it may initially appear since any arbitrary finite set of sizes may be chosen as a reasonable approximation to the continuum.

Both pure and mixed strategy models will be considered throughout. In cases of possible ambiguity, the choice will be specified.

2.2 Monomorphic ESS's and Stability

The clearest situation to understand and develop the static stability conditions of evolutionary game theory is that of a single species where all individuals initially present use the same strategy S^*, which may be either pure or mixed. The population is said to be at a monomorphic evolutionarily stable strategy (ESS), if it cannot be invaded by a small (relative to the number in the initial population) subpopulation of individuals using a different monomorphic strategy S.

Let $W(S_1, \mu)$ be the fitness of a single S_1-strategist in a population μ consisting of a few S individuals in a much larger number of S^* individuals. Intuitively, for S^* to be non-invadable by S (see Theorem 2.2.2), we need

$$W(S, \mu) < W(S^*, \mu). \tag{2.2.1}$$

Let $\mu = \varepsilon S + (1 - \varepsilon)S^*$ represent the population when there is a (small) proportion ε of S-individuals in the total population. That is, μ is the *mean strategy* of the population. For this section and most of the text, we assume that $W(S_1, \mu)$ is of the form $S_1 \cdot A\mu$ where A is the $m \times m$ *payoff matrix* and the fitness is the standard inner product in R^m of the two vectors S_1 and $A\mu$. [Note that $A\mu$ is the column vector formed by multiplying A by the transpose of μ. For notational convenience, the mathematical symbol for transpose is omitted from such column vectors.] The inner product gives the expected fitness of S_1 in a single contest against a random individual in the total population if the entries a_{kl} of A give the fitness (i.e. payoff) of a pure k-strategist π_k in a contest with π_l.

Implicit in this matrix-game form of the fitness function is the assumption that an individual may interact with itself (so that the average strategy of the contestants for S_1 remains at μ). If this is not the case, there are finite-population effects (Maynard Smith 1988; Hines and Anfossi 1990). This complication is ignored here by assuming the population is very large; that is, effectively infinite. A comparison of $W(S, \mu) = \varepsilon S \cdot AS + (1 - \varepsilon)S \cdot AS^*$ with $W(S^*, \mu) = \varepsilon S^* \cdot AS + (1 - \varepsilon)S^* \cdot AS^*$ shows clearly that (2.2.1) holds for sufficiently small positive ε if and only if S^* satisfies the following definition.

DEFINITION 2.2.1. S^* is a *monomorphic ESS* if, for all individual strategies S different from S^*,

(i) $S \cdot AS^* \leq S^* \cdot AS^*$ and $\hspace{8em}$ (2.2.2)

(ii) $S \cdot AS < S^* \cdot AS$ if there is equality in (i). $\hspace{5em}$ (2.2.3)

These mathematical conditions are often given the following verbal translation that form the biological basis (Maynard Smith 1982) for stability in ESS theory: S^* is an ESS if an arbitrary rare mutant S does no better than S^* in its most common contests against S^*; and, if it does as well in these, then it does worse than S^* in its rare contests (against itself).

The main distinction of the dynamic approach to evolutionary game theory is to insist that S^* be a stable equilibrium (i.e. to insist that S^* be noninvadable) with respect to the dynamic underlying the given model. It often turns out that the above heuristic approach is equivalent to the dynamic approach but this must be verified through some mechanism that translates fitness into a reproductive dynamic.

For instance, in this chapter we assume that individuals reproduce parthenogenetically (or the population is haploid). Then an offspring is a clone of its genetic parent and is assumed to inherit the identical strategy. Furthermore, fitness is taken as the per capita growth rate. With N_1 and N_2 representing the number of individuals using strategies S^* and S respectively at the time t, the generation-to-generation discrete dynamic and the continuous dynamic become the systems (2.2.4) and (2.2.5) respectively.

$$N_1(t + 1) = N_1(t)S^* \cdot A\mu$$
$$N_2(t + 1) = N_2(t)S \cdot A\mu \tag{2.2.4}$$

$$\dot{N}_1(t) = N_1(t)S^* \cdot A\mu$$
$$\dot{N}_2(t) = N_2(t)S \cdot A\mu \tag{2.2.5}$$

[For the discrete dynamic (2.2.4), it is assumed the entries of A are all nonnegative since they represent the expected number of offspring. For the continuous dynamic (2.2.5), the " $\cdot$ " indicates the time derivative.]

As the system evolves we are concerned with the stability of S^*. There is no expectation for a limiting population size. [In fact, if the population mean strategy evolves towards the monomorphism S^*, then the population size will grow (or decay) exponentially at a rate $S^* \cdot AS^*$.] That is, the interest of evolutionary game theory becomes the evolution of the proportion p of the invading strategy S. Specifically, we say that S cannot invade S^* if p evolves to zero under the above dynamics whenever p is initially close to zero. In the language of dynamical systems, this is expressed by saying that S^* is *locally asymptotically stable* (l.a.s.).

From the monomorphic point of view, the justification of Definition 2.2.1 for the single-species frequency-dependent haploid evolution is

THEOREM 2.2.2. S^* is a monomorphic ESS if and only if S^* is l.a.s. for either dynamics (2.2.4) or (2.2.5) and any choice of individual strategy S different from S^*.

PROOF. The proportion p of S-users is given by $p = N_2/N$ where $N = N_1 + N_2$ is the total population size (i.e. density). Also $\mu(t) = p(t)S + (1 - p(t))S^*$. Although the possibility N tends to zero does raise the question of the biological relevance of p, there is no mathematical difficulty in the frequency dynamic below. [A more thorough discussion of density effects is postponed until Chapter 5.]

In the discrete dynamic,

$$N(t+1) = N_1(t)S^* \cdot A\mu(t) + N_2(t)S \cdot A\mu(t)$$
$$= N(t)[(N_1(t)S^* + N_2(t)S)/N(t)] \cdot A\mu(t)$$
$$= N(t)\mu(t) \cdot A\mu(t).$$

Thus,

$$p(t+1) = \frac{N_2(t)S \cdot A\mu(t)}{N(t)\mu(t) \cdot A\mu(t)} = p(t)\frac{S \cdot A\mu(t)}{\mu(t) \cdot A\mu(t)}. \qquad (2.2.6)$$

Since $\mu(t) \cdot A\mu(t)$ is always positive, $\mu(t+1)$ will be closer than $\mu(t)$ is to S^* iff $p(t+1) < p(t)$ iff $S \cdot A\mu(t) < \mu(t) \cdot A\mu(t)$ for all μ sufficiently close (but not equal) to S^*. [The notation iff stands for if and only if.]

Similarly, for the continuous dynamic, $\dot{N} = N_1 S^* \cdot A\mu + N_2 S \cdot A\mu = N\mu \cdot A\mu$. Thus

$$\dot{p} = d/dt(N_2/N) = (NN_2 S \cdot A\mu - N_2 N\mu \cdot A\mu)/N^2 = p(S - \mu) \cdot A\mu. \qquad (2.2.7)$$

That is, $\dot{p} < 0$ iff $S \cdot A\mu < \mu \cdot A\mu$.

Thus, for either dynamic, p evolves to zero whenever p is initially close to zero iff $S \cdot A\mu < \mu \cdot A\mu$ for all μ sufficiently close (but not equal) to S^*. The discussion immediately before Definition 2.2.1 completes the proof.

∎

2.3 The Hawk-Dove Game

In general, monomorphic ESS's are difficult to determine since their definition depends on the other individual strategies present in the population. However, if there are exactly two pure strategies (the case of one pure strategy has no biological or mathematical interest), not only is the task simplified greatly but biologically relevant models occur.

For instance, the so-called Hawk *(H)* — Dove *(D)* game that is a simple model of aggressive versus non-aggressive behavior is one of the first and foremost examples of evolutionary game theory (Maynard Smith 1982). In this scenario, the pairwise contests are over a resource of fitness value *V.* The pure Hawk behavior escalates any confrontation until either it wins or loses the contest or else the opponent leaves. A Dove displays unless the opponent leaves and will leave itself if the opponent escalates. [It should be pointed out here that the pure strategies are for a single species and the descriptive terms hawk and dove refer only to an individual's propensity to fight.] If contestants in the average fight incur a total cost C, the 2×2 payoff matrix A is given by

$$\begin{array}{cc} & \begin{array}{cc} H & \qquad D \end{array} \\ \begin{array}{c} H \\ D \end{array} & \left[\begin{array}{cc} (1/2)(V-C) & V \\ 0 & V/2 \end{array} \right]. \end{array}$$

For instance, the H versus H entry is $(1/2)(V-C)$ since each contestant expects a gain of $V/2$ while incurring a loss of $C/2$, whereas H gains payoff V in a contest against D. If the dynamic is discrete, a constant may be added to each entry to reflect a background fitness independent of this particular resource in order that all payoffs are positive.

Individual strategies in this game all lie on the line segment Δ^2 joining $e_1 = (1, 0)$ to $e_2 = (0, 1)$ in Figure 2.3 with aggression greater for those near the Hawk strategy $(1, 0)$. If the cost of fighting is smaller than the value of the resource (i.e. $C \le V$), it pays an individual to be as hawkish as possible. In the pure-strategy model, H is then the only monomorphic ESS. Otherwise, it is the individual (mixed) strategy that is closest to H in Figure 2.3.

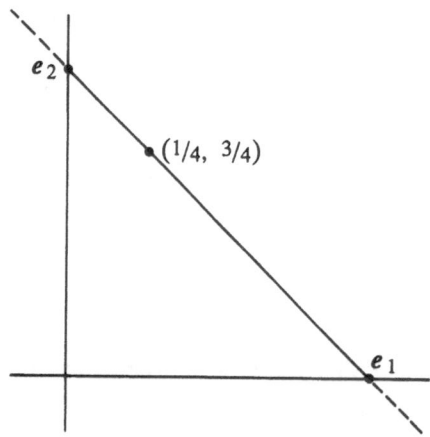

Figure 2.3. *The strategy simplex for the hawk-dove game*

The one-dimensional simplex, Δ^2, is the solid line segment between pure strategies $e_1 = (1, 0)$ and $e_2 = (0, 1)$ (i.e. H and D respectively). If $C = 4$ and $V = 1$, then $S_0^* = (1/4, 3/4)$ is the ESS. Super-strategies are points on the dashed line that extends Δ^2 to the second and fourth quadrants. For two-strategy games, Δ^2 is often sketched as a horizontal line segment by deleting the coordinate axes.

Suppose $C > V$ for the rest of this section. In the pure-strategy model, there are no monomorphic ESS's as neither a population of all H nor all D is resistant to invasion (since $H \cdot AD > D \cdot AD$ and $D \cdot AH > H \cdot AH$). On the other hand, any mixed-strategy model that includes individuals using strategy $S_0^* = (V, C - V)/C$ has S_0^* as its only ESS. Moreover, any other mixed-strategy model has a monomorphic ESS if and only if all individual mixed strategies are either between S_0^* and D or else between S_0^* and H. In such a case, the monomorphic ESS is that strategy closest to S_0^*.

From this simple example, it is clear that the existence and description of monomorphic ESS's depend critically on the initial set of individual strategies. However, the pure-strategy dynamics of (2.2.4) and (2.2.5) both induce the evolution as in (2.2.6) and (2.2.7) of the frequency p of Hawks to V/C for any initial $p \neq 0$, 1. That is, neither monomorphic pure-strategy population can resist an invasion; rather each will evolve to the polymorphic population containing Hawks and Doves with frequencies given by the components of S_0^*.

2.4 The Static Characterization of an ESS

(A) Biological Considerations

The disturbing dependence in Section 2.3 on the set of initial strategies when characterizing monomorphic ESS's and the evolution of the strategy frequency to a polymorphism leads to a more powerful interpretation of the ESS as a population characteristic as opposed to the individual characteristic used in Section 2.2. Let us consider the pure-strategy model first. Here, any combination of pure-strategies may invade the current population.

Suppose the mean strategy of the population is currently at S^*. The reader must be careful of a shift in notation here. In Section 2.2, all three probability vectors (S^*, S and u) lie in the same set Δ^m. Although mathematicians will find the use of the two different types of letters (Greek and Roman) unusual to denote elements in the same set, it was done to emphasize the biological difference between a population mean strategy (μ) and individual strategies (S and S^*). For polymorphic populations, it is the mean strategy that may resist invasion, not a particular individual strategy. Since ESS's will continue to be denoted by S^* (e.g. Definition 2.4.1), Roman letters will now be used for both. The interpretation should be clear from the context in which S is used.

In the pure-strategy model, the conditions of Definition 2.2.1 must now be checked for all invading subpopulations with mean strategy S. It is not enough to check for stability using one invading pure strategy at a time — rather stability against invasion must hold for simultaneous invasion by different pure strategies. This important distinction, pointed out by Maynard Smith (1982, Appendix D), implies the following standard conditions of a (polymorphic) ESS.

DEFINITION 2.4.1. S^* is an *ESS* if, for all $S \in \Delta^m$ different from S^*,

(i) $S \cdot AS^* \leq S^* \cdot AS^*$ and   (2.4.1)

(ii) $S \cdot AS < S^* \cdot AS$ if there is equality in (i).   (2.4.2)

There is an equivalent ESS definition for matrix games that has gained equal stature over the past decade. [The proof of equivalence (omitted) follows directly from the inequality (2.2.1) together with the compactness of Δ^m as in Hofbauer and Sigmund (Theorem 15.5, 1988).] We will choose the most convenient definition in a given situation. In particular, Definition 2.4.1 is used to determine ESS's in part B of this section while Definition 2.4.2 is used in Section 2.5 to prove dynamic stability.

DEFINITION 2.4.2. S^* is an *ESS* if

$$S^* \cdot AS > S \cdot AS$$ (2.4.3)

for all $S \neq S^*$ in some neighborhood of S^* in Δ^m.

There are differences of opinion on the ESS definition in mixed-strategy models. Suppose now that S_1, S_2, $\cdots$ S_n list all possible individual phenotypes. [Recall that we do not concern ourselves with the practical problem of how such a list can be determined in a real-life application; rather we take this list as given along with the payoff matrix.] If p_i is the frequency of individuals using S_i in the population, then $S = \sum_{i=1}^{n} p_i S_i$ is the convex combination of $\{S_1, S_2, \cdots, S_n\}$ giving the mean strategy or state. S^* must be one such convex combination.

One possible interpretation of the ESS conditions arises when we check that the conditions of Definition 2.4.1 are satisfied only for those states S in the closed convex hull $\{\sum p_i S_i \mid p_i \geq 0, \sum p_i = 1\}$ generated by the mixed strategies. That is, all of the invading subpopulations are assumed to have their mean strategy in this set. In the hawk-dove game of Section 2.3, under this alternative, the model with $V = 2$, $C = 4$ and the two mixed strategies $S_1 = (0, 1)$, $S_2 = (.2, .8)$ has S_2 as its ESS since it is the closest mixed strategy to the ESS of Definition 2.4.1; namely, $S_0^* = (2, 4-2)/4 = (1/2, 1/2)$. This leads to the concept of a conditional ESS that is useful for the analysis of genotypic equilibria encountered in the diploid models of Chapter 4.

The alternative point of view, adopted for most of this book, is that an ESS such as S_2 should not be considered evolutionarily stable; its stability is an artificial consequence of the arbitrary restriction on the set of possible individual mixed strategies. Evolution would encourage mutations that are more aggressive than either S_1 or S_2 and the population mean would shift towards $S_0^* = (1/2, 1/2)$. In particular, for single-species haploid models, we insist an ESS resist invasion by any $S \in \Delta^m$. That is, ESS's are given by Definitions 2.4.1 and 2.4.2.

An even more extreme point of view (see also Section 5.4) would question the evolutionary stability of H in Section 2.3 when $V > C$. Although the pure Hawk strategy is the only ESS by Definition 2.4.1, when this model is put in the biological context, one could argue that these interactions should encourage the appearance of mutant Hawks that are better fighters than H. For instance, when $V = 3$ and $C = 2$, the hawk-dove payoff matrix $A = \begin{bmatrix} 1/2 & 3 \\ 0 & 3/2 \end{bmatrix}$ has $S_0^* = (3/2, -1/2)$. Geometrically, this Super-Hawk (SH) strategy S_0^* is a point in the fourth quadrant on the line containing Δ^2 in Figure 2.3. If the payoff of S_1 in a contest against S_2 is still assumed to be $S_1 \cdot AS_2$, we obtain payoffs

$$
\begin{array}{c@{\quad}c@{\quad}c}
 & SH & H & D \\
\begin{array}{c} SH \\ H \\ D \end{array} &
\begin{bmatrix} -3/4 & 3/4 & 15/4 \\ -3/4 & 1/2 & 3 \\ -3/4 & 0 & 3/2 \end{bmatrix}.
\end{array}
\tag{2.4.4}
$$

Note that the relative magnitudes of the entries can be reasonably interpreted as asserting SH is a better and more ferocious fighter than H. Furthermore, SH is the only ESS of the 3×3 extended payoff matrix. Although I feel such extensions of ESS theory are biologically relevant, I will not pursue them in this text since super-strategies outside the simplex Δ^m are not allowed in the standard model of evolutionary game theory.

(B) *Mathematical Considerations*

Definition 2.4.1 (and the verbal translation following Definition 2.2.1) is often taken as the biological meaning of an ESS. This static definition also has formal mathematical advantages over Definition 2.4.2 which is more closely connected to the haploid dynamic. Indeed, the (symmetric Nash) equilibrium condition (2.4.1) that plays a central role in classical game theory has numerous mathematical consequences that are relevant to biology along with those of the stability condition (2.4.2) (Selten 1983; Hofbauer and Sigmund 1988).

One set of consequences produces a systematic method to find all ESS's of any payoff matrix (Haigh 1975). However, for the purposes of this text, the ability to understand the method as it applies to the following examples (taken from other sections of the text) will suffice. These examples also illustrate Haigh's results, Theorems 2.4.4 and 2.4.5 below. [For a more sophisticated analysis that characterizes possible patterns of ESS's, see Vickers and Cannings (1988).]

One reason ESS theory has grown rapidly in acceptance is that many practical applications involve only two pure strategies and, for these games, ESS's are easy to determine. Any $A = \begin{bmatrix} a & b \\ c & d \end{bmatrix}$ has an ESS unless there is *selective neutrality* (i.e. $a = c$ and $b = d$). If $a < c$ and $d < b$ (as was the case in the hawk-dove game with $C > V$), there is a unique ESS $S^* = \dfrac{1}{c - a + b - d}(b - d, c - a)$ and this satisfies $AS^* = \lambda(1, 1)$ where λ is the real number $(bc - ad)/(c - a + b - d)$. Otherwise, the pure strategy $e_1 = (1, 0)$ (respectively e_2) is an ESS if either $a > c$ or $a = c$ and $b < d$ (respectively, $d > b$ or $d = b$ and $a < c$).

ESS's for games with more than two pure strategies are harder to determine. The first example, with payoff matrix given in (2.4.4), has ESS e_1 since

(i) all entries in the first column are the same implying (2.4.1) is an equality for all $S \in \Delta^3$ since $Ae_1 = \lambda(1, 1, 1)$, and

(ii) $S^* \cdot AS > S \cdot AS$ since each entry in the first row (except that corresponding to e_1) is larger than the other entries in its corresponding column.

Furthermore, there is no other ESS since $Ae_1 = \lambda(1, 1, 1)$.

As a second example, the ESS structure of payoff matrix (2.7.3) is identical to that of

$$A = \begin{bmatrix} 0 & 2 & -1 \\ -1 & 0 & 2 \\ 2 & -1 & 0 \end{bmatrix} \tag{2.4.5}$$

since the addition of the constant W_0 to each entry of a column of A does not change either condition of Definition 2.4.1. $S^* = (1/3, 1/3, 1/3)$ is an ESS since

(i) $AS^* = 1/3(1, 1, 1)$ implies equality in (2.4.1), and

(ii) $(S^* - S) \cdot AS = (S^* - S) \cdot A(S - S^*) > 0$ if $S \in \Delta^3$ and
   $(x_1, x_2, x_3) = S - S^* \neq 0$ since $x_1 + x_2 + x_3 = 0$ and
   $(x_1, x_2, x_3) \cdot A(x_1, x_2, x_3) = x_1 x_2 + x_2 x_3 + x_1 x_3$
   $= 1/2[(x_1 + x_2 + x_3)^2 - (x_1^2 + x_2^2 + x_3^2)]$ is negative.

Furthermore S^* is the only ESS. [If there were another ESS T^*, then $S \cdot AT^* \leq T^* \cdot AT^*$ for all $S \in \Delta^3$. This contradicts the fact $S^* \cdot AT^* > T^* \cdot AT^*$.]

There may be more than one ESS, as seen by the payoff matrix of Table 4.4

$$A = \begin{bmatrix} 5 & 5 & 5 \\ 11 & 4 & 5 \\ 11 & 1 & 7 \end{bmatrix} \tag{2.4.6}$$

The pure strategy e_3 is an ESS since the diagonal entry is the largest in the third column. $S^* = (1/7, 6/7, 0)$ is also an ESS since $(1/7, 6/7)$ is an ESS of the 2×2 subgame $\begin{bmatrix} 5 & 5 \\ 11 & 4 \end{bmatrix}$ and the third component of $AS^* = (5, 5, 17/7)$ is smaller than the other two.

Finally, there are many 3×3 games that have no ESS. For instance, the hawk-dove game can be extended to a 3×3 game by considering $(1/2, 1/2)$ as a pure strategy (Section 6.1). The new payoff matrix (using $V = 2$ and $C = 4$) is

$$A = \begin{bmatrix} -1 & 2 & 1/2 \\ 0 & 1 & 1/2 \\ -1/2 & 3/2 & 1/2 \end{bmatrix}. \tag{2.4.7}$$

There is no ESS since $AS = \lambda(1, 1, 1)$ for any $S \in \Delta^3$ of the form $(p, p, 1 - 2p)$ and no pure strategy is an ESS.

When there are more than two pure strategies, the following abstract concepts can be used to produce a short list of possible ESS's (and also to verify the conclusions stated in the above examples). Some of these concepts prove useful for general vectors u in R^m as opposed to $S^* \in \Delta^m$.

DEFINITION 2.4.3. (a) Let A be an $m \times m$ payoff matrix and $u = (u_1, u_2, \cdots, u_m)$ be a vector in R^m. The *support* of u is given by $\text{supp}(u) = \{i \mid u_i \neq 0\}$ and the *range* of $Au = (v_1, v_2, \cdots, v_m) \in R^m$ is given by $R(u) = \{i \mid v_i = \max\{v_1, v_2, \cdots, v_m\}\}$.

(b) $X^m = \left\{ x = (x_1, x_2, \cdots, x_m) \in R^m \mid \sum\limits_{i=1}^{m} x_i = 0 \right\}$ is the subspace of R^m parallel to the simplex Δ^m.

(c) Let S be a strategy in Δ^m. S is in the *interior* of Δ^m (or S has *full support*) if $\text{supp}(S) = \{1, 2, \cdots, m\}$.

The support of an ESS $S^* \in \Delta^m$ has obvious biological importance — it specifies the pure strategies that are in use by the population. The range of AS^* is tied directly to ESS condition (2.4.1) that implies $\text{supp}(S^*) \subseteq R(S^*)$. Moreover, (2.4.2) need only be verified for those S with $\text{supp}(S) \subseteq R(S^*)$. That is, we have

THEOREM 2.4.4. S^* is an ESS if and only if $\text{supp}(S^*) \subseteq R(S^*)$ and $S \cdot AS < S^* \cdot AS$ for all $S \in \Delta^m$ different from S^* with $\text{supp}(S) \subseteq R(S^*)$.

Since $\text{supp}(S^*) \subseteq R(S^*)$, $e_i \cdot AS^*$ is the same constant λ for all $i \in \text{supp}(S^*)$. In particular, if S^* is in the interior of Δ^m, $AS^* = \lambda 1$ where $1 = (1, 1, \cdots, 1) \in R^m$. In this case, we must verify $x \cdot Ax < 0$ for all nonzero $x \in X^m$ as was done for the interior ESS of the payoff matrix (2.4.5). In general, we have

THEOREM 2.4.5. Suppose S^* is in the interior of Δ^m. The following three statements are equivalent.

(a)　S^* is an ESS for A.

(b)　If $S \neq S^*$, then $S \cdot AS^* = S^* \cdot AS^*$ and $S \cdot AS < S^* \cdot AS$.

(c)　$AS^* = \lambda 1$ (i.e. $R(S^*) = \{1, 2, \cdots, m\}$) and $x \cdot Ax < 0$ for all nonzero $x \in X^m$ (i.e. A is *negative definite* on X^m).

From Theorem 2.4.4, Haigh (1975) pointed out that the support of one ESS cannot be contained in the support of another (in fact, in the range of another). This elementary result is extremely useful. For instance, if S^* is an interior ESS, it is the only ESS. Overall, the text does not emphasize these static techniques to find ESS's. In fact, the rest of Chapter 2 is mostly concerned with the polymorphic population dynamic.

2.5 Stability for the Continuous Dynamic

This section and the next two are concerned with the dynamic consequences of Definitions 2.4.1 and 2.4.2 for polymorphic populations. Specifically, the stability of the population's mean strategy, S, is examined near an ESS, S^*.

Let p_i be the time-dependent frequency of phenotype S_i. A calculation analogous to that involved in (2.2.7) produces the continuous dynamic

$$\dot{p}_i = p_i(S_i - S) \cdot AS \qquad (2.5.1)$$

where $1 \leq i \leq n$ and $S = S(p) = \sum_{i=1}^{n} p_i S_i$ is the state at time t.

We will see dynamical systems similar to (2.5.1) throughout the text. It is actually a dynamic on $p = (p_1, \cdots, p_n) \in \Delta^n$. However, (2.5.1) does induce an evolution of the state S that remains for all time in the subsimplex of Δ^m given by the closed convex hull $\{\sum p_i S_i \mid p_i \geq 0, \sum p_i = 1\}$ of the phenotypes $\{S_1, \cdots, S_n\}$. [There is no relationship, a priori, between n and m.] Stability in dynamical systems such as (2.5.1) is usually concerned with that of p. However, for us, it is the stability of S^* according to the following definition that is important.

DEFINITION 2.5.1.　The state $S^* \in \Delta^m$ is *locally asympotically stable* (l.a.s.) for the dynamic (2.5.1) if, whenever the initial state $S(p)$ is sufficiently close to S^*, the population mean strategy evolves to S^*.

If S^* is to be stable, it must be a convex combination of $\{S_1, \cdots, S_n\}$, say $S^* = \sum p_i^* S_i$. One of the first and most important results of dynamic evolutionary game theory is then

THEOREM 2.5.2.　If an ESS, S^*, is a convex combination of the phenotypes initially present, then S^* is l.a.s. for the evolution of the mean strategy determined by (2.5.1).

PROOF. If $S(p^*) = S^*$, then p^* is an equilibrium of (2.5.1) by Theorem 2.4.4 since supp (S_i) is contained in supp (S^*) whenever $p_i^* > 0$. To show stability, let $V(p) = \Pi p_i^{p_i^*}$ where $S^* = \Sigma p_i^* S_i$ and the product is taken over those i for which $p_i^* > 0$. Then V maps p into $[0, 1]$ and has a unique global maximum at $p = p^*$ since

$$\ln(V(p)/V(p^*)) = \left(\sum_{p_i^* > 0} p_i^* \ln(p_i/p_i^*) \right) \leq \ln\left(\sum_{p_i^* > 0} p_i \right) \leq 0$$

with equality iff $p = p^*$. Also, since $p_i \neq 0$ for all t if S_i is initially present,

$$\frac{d}{dt} \ln V(p) = \sum_{p_i^* > 0} p_i^* (\dot{p}_i/p_i) = \sum_{i=1}^{n} p_i^* (S_i - S) \cdot AS = (S^* - S) \cdot AS.$$

Thus, by Definition 2.4.2, $V(p)$ is increasing for all $S(p)$ sufficiently close to S^* with equality if and only if $S(p) = S^*$. To summarize, any p^* has a neighborhood in Δ^n such that

(i) p^* is the only local maximum of $V(p)$ in this neighborhood, and
(ii) $V(p)$ is increasing with respect to time for all $S(p)$ in this neighborhood unless $S(p) = S^*$.

Thus $\lim_{t \to \infty} V(p)$ exists and, since Δ^n is compact, there is a $p^\infty \in \Delta^n$ in the limit set of $\{p(t) | t \geq 0\}$. Then $S(p^\infty) = S^*$ and the above argument applied to p^∞ in place of p^* shows that p^∞ is the only limit point (i.e. $\lim_{t \to \infty} p(t) = p^\infty$). Since $\{p \in \Delta^n | S(p) = S^*\}$ is compact, there is a neighborhood of S^* in Δ^m such that S converges to S^* if S is initially in this neighborhood.   ∎

The above proof is based on the concept of a Liapunov function that is used at other times in this text. The following definition and theorem (Hofbauer and Sigmund 1988) summarize what will be needed.

DEFINITION 2.5.3. A differentiable real-valued function $V(x)$ is a *strict Liapunov function* for the dynamic $\dot{x} = f(x)$ if $\dot{V}(x) > 0$ whenever x is not an equilibrium. It is a strict *local* Liapunov function if the inequality is true for all non-equilibrium points x in some neighborhood of an equilibrium.

THEOREM 2.5.4. If a dynamic has a strict local Liapunov function V that has an isolated local maximum at x^*, then any initial point near x^* will evolve to an equilibrium near x^*.

REMARK 2.5.5. Theorem 2.5.2 does not assert that p^* is l.a.s. In fact, $\{p | S(p) = S^*\}$ is in general the intersection of Δ^n with a hyperplane through p^* (see Sections 6.1 and 6.2). In these cases p^* is only *neutrally stable* (i.e. if p is sufficiently close to p^* initially, then p remains close to p^*).

REMARK 2.5.6. Global stability cannot be expected in Theorem 2.5.2. For instance, the payoff matrix $\begin{bmatrix} 1 & 2 \\ 0 & 4 \end{bmatrix}$ has both pure strategies as ESS's that become l.a.s. equilibria for the pure-strategy dynamic. Any initial polymorphism evolves monotonically away from the unstable interior equilibrium $(2/3, 1/3)$ towards one of these ESS's. On the other hand, if S^* is in the interior of the phenotypes initially present, then S^* is globally stable. This is true because Definition 2.4.2 then holds for all $S \in \Delta^m$ and $V(p)$ has no local maxima other than at p^*.

REMARK 2.5.7. The above proof of Theorem 2.5.2 is the generalization of that given by Hofbauer and Sigmund (Section 16.4, 1988) for the pure-strategy dynamic. As mentioned in Cressman (1990), other partial proofs can be found in Taylor and Jonker (1978) who relied on a regularity condition for the ESS (see Section 2.8 below), Zeeman (1981), Hines (1980) and Akin (1982). The latter two also considered the mixed-strategy dynamic.

2.6 The Strong Stability Concept and the Dynamic Characterization of an ESS

Almost from the first appearance of the ESS concept in Maynard Smith and Price (1973), its validity as a predictor of evolutionary stability in polymorphic populations has been questioned. Although Theorem 2.5.2 goes a long way in justifying ESS theory in haploid species, it does not settle the issue. One such doubt is the topic of this section; namely, the converse of Theorem 2.5.2.

The basic question becomes, "Are there other dynamically stable equilibria besides ESS's?" The answer is not clearcut.

For the pure-strategy dynamic with two strategies, the ESS conditions are a complete static characterization of l.a.s. equilibria. On the other hand, for three (or more) pure strategies, examples abound (e.g. Zeeman (1979) and Appendix D of Maynard Smith (1982)) of non-ESS equilibria that are l.a.s. for the pure-strategy continuous dynamic. [That is, the answer to the basic question seems to be "Yes".] As implied in the Introduction, such examples are exactly the type of situation evolutionary game theory hoped to avoid in that the evolutionary outcome of selection cannot be predicted in them without recourse to the underlying dynamic.

There are two obvious ways to attempt to rectify this situation. One is to loosen the ESS concept to include cases such as in the previous paragraph. However, there appear to be no across-the-board static conditions involving fitness comparisons that are equivalent to pure-strategy dynamic stability for general matrix games. In keeping with my view of evolutionary game theory, this text therefore pursues the other alternative; namely, retain Definition 2.4.1 of an ESS while strengthening the stability concept so that the above cases are not "strongly stable". [The reader is warned that other authors use the term strongly stable in different contexts.]

DEFINITION 2.6.1. S^* is *strongly stable* for the continuous dynamic if, whenever S^* is a convex combination of an arbitrary choice of initial phenotypes $\{S_1, S_2, \cdots, S_n\}$, S^* is a l.a.s. equilibrium for the mean strategy evolution determined by (2.5.1).

By Theorem 2.5.2, an ESS is strongly stable. Conversely, if S^* is strongly stable, it must be a monomorphic ESS as in Definition 2.2.1 by considering $\{S^*, S\}$ as the choice of initial phenotypes where $S \in \Delta^m$ is arbitrary. Thus any strongly stable S^* is an ESS as given by Definition 2.4.1. That is, ESS's are characterized dynamically by

THEOREM 2.6.2. S^* is an ESS if and only if S^* is strongly stable for the continuous dynamic.

Thus, the examples of non-ESS l.a.s. equilibria referred to above must depend critically on the fact pure strategies were the only choices considered for the initial phenotypes. Indeed, Maynard Smith (1982) explicitly provides another choice of mixed strategies for which his non-ESS equilibrium becomes unstable. To reiterate Theorem 2.6.2, for any non-ESS equilibrium, it is possible to pick phenotypes $\{S_1, S_2, \cdots, S_n\}$, such that the equilibrium is a convex combination of them and, at the same time, the dynamic (2.5.1) is not l.a.s.

An unexpected, though pleasant, consequence of this strengthened stability concept is that it appears to closely reflect the original heuristic that an ESS should be noninvadable by any combination of pure or mixed mutant strategies. My own opinion is that strong stability and Theorem 2.6.2 constitute the ideal to which all theoretical models in evolutionary game theory should strive because they give both a clear intuitive understanding of the static conditions and a concise statement of the dynamic consequences. This partnership between intuition and analysis becomes extremely powerful in applications of the theory.

The topics of Chapter 2 that deal with the continuous-dynamic, frequency-dependent, single-haploid-species model can be rearranged to fit the following four-step ESS program. [The brackets in each step refer to the relevant part of Chapter 2.]

STEP 1.   Develop the frequency-dependent dynamic for the invasion of one monomorphism, S^*, by another, S. (Equation (2.2.7))

STEP 2.   Find static criteria involving fitness comparisons that are equivalent to monomorphic dynamic stability. (Definition 2.2.1 and Theorem 2.2.2)

STEP 3.   Define an ESS as a polymorphic population characteristic that satisfies the criteria from Step 2 for all other choices of S. (Definition 2.4.1)

STEP 4.   Develop the general polymorphic frequency-dependent dynamic (Equation (2.5.1))
and characterize the ensuing dynamic stability of an ESS. (Theorems 2.5.2 and 2.6.2)

To a large extent, the text can be viewed as an implementation of this four-step program to
other models in evolutionary biology (see Introduction). Specifically, Chapters 3 and 5 succeed
in varying degrees to generalize the program to frequency-dependent two species and density-
dependent single species models respectively.

2.7 Stability for the Discrete Dynamic

Biologists often use ESS theory to explain the behavioral outcome of evolution for observed
species and to predict how these same species will behave in different environments (i.e. subject
to different payoffs). Sections 2.5 and 2.6 then provide strong evidence to conclude all relevant
stable evolutionary outcomes are among the ESS's of the given models. The theory from these
sections is based entirely on the continuous dynamic. On the other hand, many researchers
consider the discrete dynamic to be more realistic (e.g. individuals need not reach reproductive
maturity instantaneously). Even though the dynamic (2.5.1) can be viewed as a continuous
approximation of the discrete version (2.7.1), examples in this section will caution us not to
automatically transfer stability results from one dynamic to the other.

Clearly, by Theorem 2.2.2, Steps 1 and 2 above imply a strongly stable equilibrium, S^*, for
the polymorphic discrete dynamic

$$p_i(t + 1) = p_i(t) \; \frac{S_i \cdot AS}{S \cdot AS} \tag{2.7.1}$$

must satisfy the conditions of Definition 2.4.1. That is, S^* must be an ESS.

Unfortunately, these conditions are no longer sufficient to guarantee stability as illustrated by
the example in Taylor and Jonker (1978) of an unstable ESS in the three pure-strategy discrete
dynamic with payoff matrix

$$\begin{bmatrix} -\varepsilon & 1 & -1 \\ -1 & -\varepsilon & 1 \\ 1 & -1 & -\varepsilon \end{bmatrix}. \tag{2.7.2}$$

For any $\varepsilon > 0$, $S^* = (1/3, 1/3, 1/3)$ is an ESS that is unstable under (2.7.1). As they
explain, while the continuous dynamic in this example spirals slowly inwards towards S^*, the
discrete analogue follows the tangent to the trajectory far enough at each point that an outward
spiral results.

Thus, Step 4 of the program set out in the last section has failed for the polymorphic discrete
dynamic. A pessimistic conclusion is that the program is wrong and other means should be

sought to generate static stability conditions. The remainder of this section is primarily intended to support another (more optimistic?) conclusion that a discrete dynamic is, a priori, incompatible with theoretical evolutionary games. However, it should first be noted there is compatibility when there are at most two pure strategies. In this case, the population mean strategy converges monotonically towards an ESS under (2.7.1).

Suppose, for now, that static stability conditions do exist. The counterexample from (2.7.2) creates the opposite problem to that discussed in Section 2.6 — now either the ESS conditions need to be strengthened or the strong stability concept watered down. The difficulties this leads to when there are more than two pure strategies are perhaps better illustrated by the example taken from Rowe et al. (1985) with payoff matrix

$$\begin{bmatrix} W_0 & W_0 + 2 & W_0 - 1 \\ W_0 - 1 & W_0 & W_0 + 2 \\ W_0 + 2 & W_0 - 1 & W_0 \end{bmatrix}. \tag{2.7.3}$$

The constant, W_0, may be regarded biologically as a strategy-independent background fitness that is shared by all individuals in the population. [For a more detailed discussion of the dynamic consequences of such a constant, see Rowe et al. (1985) or Cressman et al. (1986).]

As pointed out by Rowe, $S^* = (1/3, 1/3, 1/3)$ is an unstable ESS when $W_0 = 1$ and stable when $W_0 = 3$ for the pure-strategy dynamic (2.7.1). [Also note that, unlike (2.7.2), all entries of (2.7.3) are nonnegative for these choices of W_0. That is, these matrices are legitimate choices for the discrete dynamic.] On the other hand, static comparisons of strategy fitnesses, the cornerstone of evolutionary game theory, cannot produce different conditions for different values of W_0.

The resulting inconsistency of static versus dynamic stability concepts seems impossible to avoid and led Rowe to conclude correctly that the stability concept of discrete evolutionary game theory must ultimately be based on the particular dynamic. This is not to suggest that the ESS of Definition 2.4.1 has no dynamic role whatsoever in discrete models. For instance, it seems an ESS which is l.a.s. for the pure-strategy discrete model may remain so for all related mixed-strategy situations (personal discussions with G. Vickers, 1990) — a result definitely not true for non-ESS equilibria.

At the same time, the practitioner of applied ESS theory must be cautioned that the precise discrete-dynamic stability properties of these equilibria are not well-understood. For this reason the text relies almost exclusively on continuous models from here on.

2.8 Alternative Proof of Theorem 2.5.2

In Section 2.5, the first important theorem of evolutionary game theory was proved by means of a specific local Liapunov function. In fact, the original proof of this theorem (in the pure-strategy model) by Taylor and Jonker (1978) used a linearization method similar to that of this section. They assumed a regularity condition to ensure no eigenvalue of the linearized system at the ESS, S^*, has zero real part. Since the Liapunov approach avoids this difficulty and can also be used effectively for mixed-strategy models, it has become the preferred method to prove Theorem 2.5.2.

The purpose of this section is twofold. The first is to demonstrate that the linearization approach, although indisputably more difficult for the model of this chapter, produces the same important result; namely, Theorem 2.5.2. The other purpose, perhaps more important, is to lay the groundwork for the local stability analysis of Chapters 3, 4 and 5 where no explicit Liapunov functions are readily available for the general nonlinear dynamic.

The method relies on the following facts. S^* will be l.a.s. if all eigenvalues of the $n \times n$ matrix, given by the linear terms of (2.5.1) expanded about S^*, have negative real part but unstable if some eigenvalues have positive real part. If there are eigenvalues of zero real part (i.e. the equilibrium is not *hyperbolic*), a more thorough analysis of higher order terms using centre manifold theory determines l.a.s.

To efficiently handle the problem of eigenvalues with zero real part, it seems best to include a selectively-neutral density effect in the continuous polymorphic dynamic analogous to (2.2.5); namely,

$$\dot{N}_i = N_i (S_i \cdot AS + \rho(N)) \tag{2.8.1}$$

where $i = 1, 2, \cdots, n$ and $\rho(N)$ is a smooth (i.e. continuously differentiable) real-valued strictly-decreasing function of the density $N = \Sigma N_i$. Let us postpone the biological interpretation of this density effect until Chapter 5 where general density dependence is considered. For now, $\rho(N)$ can be thought of as a background fitness that is independent of an individual's strategy and that decreases as population size increases.

From (2.8.1), $\dot{N} = \Sigma \dot{N}_i = N(S \cdot AS + \rho(N))$ and, if $N \neq 0$,

$$\dot{p}_i = \frac{d}{dt} (N_i/N) = \frac{N_i(S_i \cdot AS + \rho(N))N - N_iN(S \cdot AS + \rho(N))}{N^2}$$

$$= p_i(S_i - S) \cdot AS.$$

That is, the frequency dynamic is the same as (2.5.1) and is unaffected by the neutral density term. From this, and the fact $\rho(N)$ is strictly decreasing, it is not hard to show (proof omitted) the following.

LEMMA 2.8.1. Suppose (S^*, N^*) is an equilibrium of (2.8.1) with N^* positive. Then (S^*, N^*) is l.a.s. if and only if S^* is a l.a.s. equilibrium of (2.5.1).

To analyze the stability of (2.8.1) at (S^*, N^*) by linearization techniques, it is convenient mathematically to include the constant $\rho(N^*)$ in all entries of A. This has no effect on the dynamic but does imply $\rho(N^*) = 0$ and $S^* \cdot AS^* = 0$. In fact, it will also be assumed that $N^* = 1$ in what follows. Then $S^* = \sum_{j=1}^{n} p_j^* S_j$ where $p_j^* = N_j^*$. Let $N_i = p_i^* + x_i$. Then,

in terms of x_i, $\quad S = \dfrac{1}{N} \Sigma (p_j^* + x_j) S_j = \dfrac{1}{N} \left(S^* + \sum_{j=1}^{n} x_j S_j \right)$ and $N = \Sigma N_j = 1 + \Sigma x_j$.

The linearization of (2.8.1) is

$$\dot{x}_i = (N_i^* + x_i) [S_i \cdot A ((S^* + \Sigma x_j S_j)/(1 + \Sigma x_j)) + \rho(1 + \Sigma x_j)]$$
$$= p_i^* [S_i \cdot AS^* (1 - \Sigma x_j) + S_i \cdot A \Sigma x_j S_j + \rho'(N^*) \Sigma x_j)] \qquad (2.8.2)$$
$$+ x_i [S_i \cdot AS^*] + o(|x|)$$

where $\rho'(N^*) = \dfrac{d\rho}{dN}(N^*)$ and $o(|x|)$ tends to zero faster than $|x| = (\Sigma x_i^2)^{1/2}$ does.

At an ESS, S^*, we have $S_i \cdot AS^* \le S^* \cdot AS^* = 0$ with equality if $p_i^* > 0$. The linear terms of (2.8.2) have the form $\dot{x}_i = \Sigma_j L_{ij} x_j$ where

$$L_{ij} = \begin{cases} \delta_{ij} S_i \cdot AS^* & \text{if } p_i^* = 0 \\[2ex] p_i^* [S_i \cdot AS_j + \rho'(N^*)] & \text{if } p_i^* \ne 0 \end{cases} \qquad (2.8.3)$$

and δ_{ij} is the Kronecker delta function that is 1 when $i = j$ and 0 otherwise. Reorder the n phenotypes so that $p_i^* = 0$ if and only if $i = 1, \cdots, k$ where $0 \le k < n$. Then L has the block form

$$\begin{bmatrix} D & O \\ C & B \end{bmatrix} \qquad (2.8.4)$$

where D is the $k \times k$ diagonal matrix with nonpositive entries, B is a square matrix of order $(n - k)$, O is the zero matrix and C is unimportant (for now) when discussing eigenvalues.

LEMMA 2.8.2. If $\rho'(N^*) < 0$, then all nonzero eigenvalues of B have negative real part at an ESS, S^*. Moreover, the zero eigenspace of B is exactly equal to the subspace given by $\{x = (0, 0, \cdots, 0, x_{k+1}, \cdots, x_n) \in R^n | \Sigma x_i S_i = 0\}$. This subspace is parallel to the hyperplane $\{p = (0, 0, \cdots, 0, p_{k+1}, \cdots, p_n) | \Sigma p_j S_j = \Sigma p_j^* S_j\}$.

PROOF. Since $(Lp^*)_i = \sum_{j=1}^{n} L_{ij} p_j^* = \sum_{j=k+1}^{n} p_i^* [S_i \cdot AS_j + \rho'(1)] p_j^* = p_i^* \rho'(1)$, we have that $(0, 0, \cdots, 0, p_{k+1}^*, \cdots, p_n^*)$ is an eigenvector of L with eigenvalue $\rho'(1)$. Thus, we can restrict B to the subspace $X = \{(0, \cdots, 0, x_{k+1}, \cdots, x_n) | \Sigma x_j = 0\}$. Introduce the Shahshahani inner product (Akin 1979) on X given by $\langle x, y \rangle = \sum_{j=k+1}^{n} \frac{x_j y_j}{p_j^*}$. Then B is negative semi-definite on X since

$$\langle x, Bx \rangle = \sum_{ij} \frac{x_i p_i^* [S_i \cdot AS_j + \rho'(1)]x_j}{p_i^*}$$

$$= \sum_i x_i S_i \cdot A\Sigma x_j S_j.$$

The generalization of Theorem 2.4.5(c) shows $\langle x, Bx \rangle \le 0$ for all $x \in X$ with equality iff $\Sigma x_j S_j = 0$.

A standard result of linear algebra now proves all eigenvalues of B have nonpositive real part. [For instance, if λ is a real eigenvalue, then $Bx = \lambda x$ for some nonzero $x \in X$. Then $\langle x, x \rangle > 0$ and $\lambda\langle x, x \rangle = \langle x, Bx \rangle \le 0$ implies $\lambda \le 0$.] Furthermore, zero is the only eigenvalue that has zero real part and its corresponding eigenspace is the subspace given by $\{x | Bx = 0\} = \{x | \Sigma x_j S_j = 0\}$. This subspace is parallel to the given hyperplane since $\Sigma p_j S_j = \Sigma p_j^* S_j + \Sigma x_j S_j$ equals S^* iff $\Sigma x_j S_j = 0$. ∎

THEOREM 2.8.3. (Taylor and Jonker 1978). If S^* is an ESS that is *regular* (i.e. $S_i \cdot AS^* < S^* \cdot AS^*$ whenever $p_i^* = 0$), then S^* is l.a.s. for the pure-strategy dynamic (2.5.1).

PROOF. For the pure-strategy dynamic, Lemma 2.8.2 implies all eigenvalues of B have negative real part if the density is $\rho(N) = 1 - N$. By regularity, all eigenvalues of D are negative. Thus, (S^*, N^*) is l.a.s. and the result follows from Lemma 2.8.1. ∎

The linearization of (2.8.1) may not be sufficient to prove l.a.s. of (S^*, N^*) for a mixed-strategy model or when S^* is not regular. Since the only eigenvalue of L with nonnegative real part is zero, the stability will be determined by the flow on an invariant centre manifold of dimension l that is tangent to the zero eigenspace at (S^*, N^*). [In general, this centre manifold may not be unique. In such cases, analysis on any smooth choice will suffice for our purposes. For this reason, we will call it *the* centre manifold.]

To gain a better understanding of how centre manifold theory (Carr 1981) applies to our situation, two dynamics are analyzed to illustrate the essentially different ways L may have zero eigenvalues. The payoff matrices for these were chosen for clarity of mathematical exposition and not for biological relevance.

(A) The Centre Manifold for the Pure-Strategy Dynamic

For the first example, consider the pure-strategy dynamic with $\rho(N) = 1 - N$ and

$$A = \begin{bmatrix} -1 & 0 & 1 & -1 \\ 2 & 3 & -1 & -1 \\ 0 & 4 & -1 & 1 \\ 0 & 2 & 1 & -1 \end{bmatrix}$$

that has an equilibrium at $N^* = 1$ and $S^* = (0, 0, 1/2, 1/2)$. It is left to the reader to verify S^* is a (non-regular) ESS. From (2.8.3),

$$L = \begin{bmatrix} 0 & 0 & 0 & 0 \\ 0 & -1 & 0 & 0 \\ -1/2 & 3/2 & -1 & 0 \\ -1/2 & 1/2 & 0 & -1 \end{bmatrix}.$$

From (2.8.4), B is negative definite and D has a zero eigenvalue.

The one-dimensional centre manifold is tangent to the zero eigenvector $(2, 0, -1, -1)$. The 3-dimensional manifold $M = \{x \in R^4 \,|\, x_1 \geq 0, \; x_2 = 0\}$ is invariant under (2.8.1) and contains the centre manifold. The linearization (2.8.2) on M can be extended to include certain quadratic terms; namely,

$$\dot{x}_1 = x_1[e_1 \cdot A \, \Sigma x_j e_j + \rho'(1)\Sigma x_j + o(|x|)]$$

$$\dot{x}_3 = p_3^*[e_3 \cdot A \, \Sigma x_j e_j + \rho'(1)\Sigma x_j] + o(|x|) \qquad (2.8.5)$$

$$\dot{x}_4 = p_4^*[e_4 \cdot A \, \Sigma x_j e_j + \rho'(1)\Sigma x_j] + o(|x|).$$

By Theorem 2.10.2, it is enough to show l.a.s. on M. Define the function $V(x)$ on M by $V(x) = x_1 + 1/2(x_3^2/p_3^* + x_4^2/p_4^*) = x_1 + x_3^2 + x_4^2$. From (2.8.5),

$$\dot{V}(x) = \dot{x}_1 + 2x_3\dot{x}_3 + 2x_4\dot{x}_4$$
$$= x \cdot Ax + \rho'(1)\Sigma x_i \, \Sigma x_j + o(|x|^2). \qquad (2.8.6)$$

With $\rho'(1) = -1$, the quadratic terms of $\dot{V}(x)$ on M are

$$x \cdot Ax - (\Sigma x_i)^2 = -x_1^2 - x_3^2 - x_4^2 + x_1x_3 - x_1x_4 + 2x_3x_4 - (x_1 + x_3 + x_4)^2$$
$$= -2x_1^2 - 2x_3^2 - 2x_4^2 - x_1x_3 - 3x_1x_4$$
$$= -1/2(x_1 + x_3)^2 - 3/2(x_1 + x_4)^2 - 3/2x_3^2 - 1/2x_4^2.$$

Thus, for $x \in M$ with $|x|$ sufficiently small, $\dot{V}(x) \leq 0$ with equality if and only if x_1, x_3 and x_4 are all zero. By Section 2.5, $-V(x)$ is a strict local Liapunov function on M and so x tends to 0. Local asymptotic stability on M implies (S^*, N^*) is l.a.s. by Theorem 2.10.2.

It is interesting to note here that the magnitude of $\rho'(1)$ is important to guarantee $\dot{V}(x) \leq 0$ (even though Lemma 2.8.1 is true for any $\rho'(1) < 0$). In fact, it is the linear independence of the individual strategies that implies some such $\rho'(1)$ exists for a general pure-strategy dynamic where the centre manifold is automatically contained in the manifold

$$M = \{x \in R^n \,|\, x_1 \geq 0, \,\cdots\, , x_l \geq 0, \, x_{l+1} = 0, \,\cdots\, , x_k = 0\}$$

when $p_1^* = \cdots = p_k^* = 0$ and $S_1 \cdot AS^* = \cdots = S_l \cdot AS^* = 0$. The total mutant part of the population (i.e. $x_1 + x_2 + \cdots + x_l$) is then shown to tend to zero in M if this part is initially rare. This evolution is not necessarily monotonic but nevertheless occurs at the same time as the frequencies of strategies S_i in the support of S^* approach $p_i^* > 0$.

(B) The Centre Manifold for the Mixed-Strategy Dynamic

From Lemma 2.8.2, centre manifolds will also appear whenever the set of mixed strategies $\{S_{k+1}, \cdots, S_n\}$ is not linearly independent. For the second example, let $A = \begin{bmatrix} -1 & 1 \\ 1 & -1 \end{bmatrix}$, $S_1 = (1, 0)$, $S_2 = (0, 1)$, $S_3 = (1/2, 1/2)$ and $\rho(N) = 1 - N$. The dynamic (2.8.1) has an equilibrium when $p_1^* = p_2^* = p_3^* = 1/3$ and $N^* = 1$. The corresponding mean strategy is the ESS, $S^* = S_3$. The linearization is

$$L = 1/3 \begin{bmatrix} -2 & 0 & -1 \\ 0 & -2 & -1 \\ -1 & -1 & -1 \end{bmatrix} \tag{2.8.7}$$

where L is negative semi-definite with eigenvalues $-2/3$, -1 and 0. The zero eigenspace is spanned by the vector $(1, 1, -2)$. Since this subspace also happens to be the set $\{x \in R^3 \,|\, \Sigma x_j S_j = 0\} = \{x \,|\, x_1 = x_2, \, \Sigma x_j = 0\}$ of equilibria of (2.8.1), the entire centre manifold consists of equilibria. By Theorem 2.10.4, the zero solution of (2.8.2) is neutrally stable and any solution of (2.8.2) that starts sufficiently close to (0, 0, 0) will evolve towards a unique point in the centre manifold. Since all these points correspond to the ESS strategy, S^* will be l.a.s.

The above method generalizes to any mixed-strategy case where D from (2.8.4) has no zero eigenvalue in that the local evolution of (2.8.1) will tend towards the centre manifold consisting of a subspace of equilibrium points all corresponding to S^*. If D has zero as an eigenvalue and there is an equilibrium subspace, the general centre manifold approach becomes (substantially) more difficult since the quadratic terms in (2.8.6) do not necessarily dominate $o(|x|^2)$ near this subspace.

A particularly simple case to see this complication is the equilibrium that has $p_3^* = 1$ and $p_1^* = p_2^* = 0$ in the second example. The linearization (2.8.7) becomes

$$L = \begin{bmatrix} 0 & 0 & 0 \\ 0 & 0 & 0 \\ -1 & -1 & -1 \end{bmatrix}$$

with two-dimensional centre manifold $M = \{x \in R^3 \,|\, x_1 = x_2 \geq 0\}$. $V(x)$ from (2.8.6) is not a Liapunov function on M since

$$\dot{V}(x) = \frac{d}{dt}(x_1 + x_2 + 1/2\,x_3^2) = [x_1 + x_2 + x_3(1 + x_3)]\rho\,(1 + x_1 + x_2 + x_3)$$

$$= -(x_1 + x_2 + x_3 + x_3^2)(x_1 + x_2 + x_3)$$

is positive for those $x \in M$ that have Σx_i negative and near zero. In this case, we can use the substitute Liapunov function $V(x) \equiv \Sigma x_i - \ln(1 + x_3)$.

In the general case, the substitute Liapunov function can be obtained by considering the non-linear terms of (2.8.2). Specifically, with $\rho'(1) = -1$, we have

$$\dot{x}_i = (N_i^* + x_i)\left[\frac{1}{N}\,S_i \cdot A(S^* + \Sigma x_j S_j) - \Sigma x_j\right].$$

The problem with $V(x) = \sum_{i=1}^{l} x_i + 1/2 \sum_{i=k+1}^{n} x_i^2/N_i^*$ is that $\dot{V}(x)$ includes the unwanted cubic

term $\sum_{i=k+1}^{n} \frac{x_i^2}{N_i^*}[S_i \cdot A\Sigma x_j S_j - \Sigma x_j]$ as well as $\frac{1}{N}\Sigma x_i S_i \cdot A\Sigma x_j S_j - \Sigma x_i \Sigma x_j$. The

cubic expression can be eliminated by modifying $V(x)$ to

$$\sum_{i=1}^{l} x_i + \sum_{i=k+1}^{n}\left(1/2\,\frac{x_i^2}{N_i^*} - 1/3\,\frac{x_i^3}{N_i^{*2}}\right).$$

However, the derivative will still have an extra term; namely, $\sum \frac{x_i^3}{N_i^{*2}}[S_i \cdot A\Sigma x_j S_j - \Sigma x_j]$.

Fortunately, the process can be continued to produce the Liapunov function

$$V(x) = \sum_{i=1}^{l} x_i + \sum_{i=k+1}^{n} N_i^*\left(\sum_{j=2}^{\infty}\frac{(-1)^j}{j}\left(\frac{x_i}{N_i^*}\right)^j\right).$$

For $|x|$ sufficiently close to zero, the infinite series converges so that $V(x)$ can be written as

$$\sum_{i=1}^{l} x_i + \sum_{i=k+1}^{n}\left(x_i - N_i^*\ln\left(1 + \frac{x_i}{N_i^*}\right)\right) = \Sigma x_j - \ln \prod_{i=k+1}^{n} N_i^{N_i^*} + \Sigma N_i^*\ln N_i^*.$$ Although the

final details to show l.a.s. are omitted, the reader should note the similarity between this form of $V(x)$ and the Liapunov function $\prod p_i^{p_i}$ used in Section 2.5.

2.9 Nonlinear Fitness Functions

In all models examined so far, the dynamic has been based on the assumption that the fitness $W(S, \mu)$ of an individual S depends linearly both on the frequency distribution μ of the pure strategies exhibited by the population at a particular time and on the components of the (mixed) strategy S. Such a $W(S, \mu)$, represented as the bilinear form $S \cdot A\mu$ with payoff matrix A, is often interpreted from the viewpoint of non-cooperative game theory; that is, as the expected fitness of the individual engaged in a single contest (against a randomly chosen opponent). Most early animal behavior models (e.g. the hawk-dove game of Section 2.3) in evolutionary game theory were considered in this light.

However, evolutionary game theory applies to many situations where fitness is not a result of direct contests between individuals but of "playing the field" where fitness depends on the proportion of strategy types in the population. For instance, the hawk-dove game may be closer to reality if the fitness $F_D(p)$ of the Dove is an increasing function (of the frequency p of Doves) that remains close to zero when at least half the population is Hawk since these Hawks may be able to control all resources. In contrast, $F_D(p)$ was the linear function $1/2 \, pV$ in Section 2.3.

In general, assume that an individual using the ith pure strategy in a population with mean strategy $\mu \in \Delta^m$ has fitness $F_i(\mu)$ where F_i is smooth. Continue to take the fitness of an individual using mixed strategy S as the expected payoff $S \cdot F(\mu)$ where $F(\mu)$ is now the vector-valued frequency-dependent fitness function $(F_1(\mu), \cdots, F_m(\mu))$. If p_i is the frequency of strategy S_i, the continuous dynamic analogous to (2.5.1) becomes

$$\dot{p}_i = p_i(S_i - \mu) \cdot F(\mu) \qquad (2.9.1)$$

where $i = 1, ..., n$ and $\mu = \Sigma p_i S_i$. [Note that we will continue to use μ to denote the population mean strategy (rather than S) until ESS's are defined for these non-matrix games in part B of this section.]

The rest of this section analyzes an important example where nonlinear fitness functions arise naturally before developing the ESS program based on the dynamic (2.9.1).

(A) The Sex-Ratio Game

The explanation for the prevalence of equal numbers of (mature) males and females in animal species is usually based on the game theory principle that, if this were not the case, individual males would do well in a species where there were more females and vice versa. Maynard Smith (1982) gave a convincing argument that the sex ratio r (i.e. frequency of males in the population) is evolutionarily stable when $r = 1/2$. In the following simple model, fitness is a priori nonlinear since it is 0 (i.e. there are no offspring) at both pure-strategy extremes ($r = 0$ or $r = 1$).

Suppose the sex ratio of the offspring is determined by the female who chooses her mate at random, produces k offspring and passes the same sex ratio on to her daughters. Furthermore, suppose originally there are N females and the average sex ratio, $\dfrac{1}{N} \displaystyle\sum_{i=1}^{N} r_i$, is $\bar{r}$. If the fitness of a female using sex ratio r is measured by the number of grandchildren, then

$$W(r, \bar{r}) = k^2 \left[1 - r + r \, \frac{1 - \bar{r}}{\bar{r}} \right] \tag{2.9.2}$$

when N is large enough to ignore the possibility that siblings mate.

To verify (2.9.2), note that there are Nk children altogether, $\bar{r}Nk$ of whom are male and $(1 - \bar{r})Nk$ female. A female using sex ratio r (called female r) produces rk sons and $(1 - r)k$ daughters. These daughters will produce $(1 - r)k^2$ grandchildren whereas the sons contribute another $\dfrac{rk}{\bar{r}Nk} (1 - \bar{r})Nk^2$ since these males will mate $\dfrac{rk}{\bar{r}Nk} (1 - \bar{r})Nk$ females altogether. Thus, female r will have a grand total of $(1 - r)k^2 + \dfrac{r}{\bar{r}} (1 - \bar{r})k^2 = W(r, \bar{r})$ grandchildren. [The observant reader will have noticed that (2.9.2) is based implicitly on a discrete dynamic — something to be avoided after Section 2.7. To skirt this issue we can either appeal to the fact that there are only two pure strategies or else assert that the time between generations is negligible so that the continuous-dynamic approximation is appropriate.]

To analyze the dynamic (2.9.1) in this example, regard female r as adopting the mixed strategy $R = (r, 1 - r)$. Her fitness, (2.9.2), then has the form $R \cdot F(\bar{R})$ where

$$F(\bar{R}) = F(\bar{r}, 1 - \bar{r}) = k^2 \left(\frac{(1 - \bar{r})}{\bar{r}}, 1 \right). \tag{2.9.3}$$

The population has an average growth rate of $\bar{R} \cdot F(\bar{R}) = 2k^2(1 - \bar{r})$. If $\bar{R} \neq (1/2, 1/2)$, the population can be successfully invaded by females using strategy $(1/2, 1/2)$ because $(1/2, 1/2) \cdot F(\bar{R}) = \dfrac{k^2}{2\bar{r}} > 2k^2(1 - \bar{r})$. Thus, although the population would grow fastest if there were very few males (i.e. $\bar{r}$ close to 0), such a population would not be evolutionarily stable since females with a higher sex ratio would do well initially.

Qualitatively, the dynamic is the same as in Section 2.3 in that a monomorphic population with all females at $r^* = 1/2$ will be l.a.s. (in fact, globally stable) under (2.9.1). Furthermore, any other monomorphic population with mean strategy $(\bar{r}, 1 - \bar{r})$ will be l.a.s. if and only if invading monomorphisms $(r, 1 - r)$ are even more extreme in the sense that r is more biased than $\bar{r}$ in favor of the same sex. These same dynamic conclusions remain valid in polymorphic situations where $\bar{r}$ characterizes the average population sex ratio.

(B) ESS's and Stability

To develop ESS theory for nonlinear fitness functions in analogy to Section 2.2, suppose the monomorphic population S^* is invaded by another monomorphism S. If the frequency of S-users is to decrease under (2.9.1), we must have

$$S \cdot F(\mu) < S^* \cdot F(\mu) \qquad (2.9.4)$$

where μ is the population $\varepsilon S + (1 - \varepsilon)S^*$ consisting of a few S individuals in a much larger group of S^* individuals. But (2.9.4) is true iff $\mu \cdot F(\mu) < S^* \cdot F(\mu)$. Thus, if invasion by any monomorphism in Δ^m is possible, an ESS must satisfy the following definition (this is where we switch back to Roman letters for mean strategies).

DEFINITION 2.9.1. S^* is an *ESS* for the fitness function $F(S)$ if

$$S^* \cdot F(S) > S \cdot F(S)$$

for all $S \neq S^*$ in some neighborhood of S^* in Δ^m.

One is convinced this is the correct definition since the l.a.s. of such an ESS for the polymorphic dynamic (2.9.1) is clear by repeating the proof of Theorem 2.5.2. In fact, the above discussion is best summarized by the following version of Theorem 2.6.2.

THEOREM 2.9.2. S^* is strongly stable for the continuous dynamic (2.9.1) if and only if S^* is an ESS by Definition 2.9.1.

These considerations further promote the characterization of ESS's in matrix games as given in Definition 2.4.2. On the other hand, the more conventional heuristic of Definition 2.4.1 can be re-introduced by considering the Taylor expansion of the nonlinear fitness functions $F_i(S)$ about an equilibrium S^*. [This technique becomes even more important when stability is analyzed by linearization techniques in the next three chapters.] The Taylor expansion is

$$F_i(S) = F_i(S^*) + \nabla F_i \cdot (S - S^*) + o(|S - S^*|) \qquad (2.9.5)$$

for $i = 1, \cdots, m$ where $o(|S - S^*|)$ are nonlinear terms that approach zero quadratically as S approaches S^*. Also, ∇F_i is the gradient vector $\left(\dfrac{\partial F_i}{\partial x_1}, \cdots, \dfrac{\partial F_i}{\partial x_m} \right) \in R^m$ where all partial derivatives are evaluated at S^* and $F_i(x_1, \cdots, x_m)$ is a smooth extension to $(x_1, \cdots, x_m) \in R^m$ that agrees with $F_i(S)$ on Δ^m. [For example, the fitness functions $F_1(x_1, x_2)$ and $F_2(x_1, x_2)$ in (2.9.3) for the sex-ratio game can be taken as $k^2 \dfrac{x_2}{x_1}$ and k^2 respectively.]

If the nonlinear terms of (2.9.5) are ignored, $F_i(S)$ takes the form $(A^*S)_i$ where A^* is the $m \times m$ matrix with entries

$$A_{ij}^* = \left. \frac{\partial F_i}{\partial x_j} \right|_{S^*} + F_i(S^*) - \nabla F_i \cdot S^*. \qquad (2.9.6)$$

A^* can be thought of as a local payoff matrix but it must be emphasized that A^* can only be determined after S^* is found as opposed to the rest of the chapter where A is often used to determine the ESS through Theorems 2.4.4 and 2.4.5.

For the sex-ratio game, the equilibrium $S^* = (1/2, \; 1/2)$ has $\nabla F_1 = (-2k^2, \; 2k^2)$ and $\nabla F_2 = (0, \; 0)$. Thus A^* is the 2×2 matrix $k^2 \begin{bmatrix} -1 & 3 \\ 1 & 1 \end{bmatrix}$ that has S^* as its ESS by Definition 2.4.1. That is, l.a.s. at S^* can be shown by means of the local payoff matrix A^*. In fact, in general, we have

THEOREM 2.9.3. If S^* is an equilibrium for (2.9.1), then S^* is an ESS (according to Definition 2.9.1) if it is an ESS with respect to Definition 2.4.1 for the local payoff matrix A^* given by (2.9.6).

Although the converse of this theorem is not true (e.g. stability will depend on higher order terms in (2.9.5) if A^* is neutrally stable), the heuristic attached to the payoff matrix approach provides valuable insight into ESS theory. The method to examine stability at nonlinear ESS's suggested by Maynard Smith (1982) in his discussion of the sex-ratio game is equivalent to the determination of A^* but without this additional insight.

2.10 Appendix on Centre Manifold Theory

Centre manifold theory can be used to characterize the stability of an equilibrium point of a dynamical system when the linearized dynamic has eigenvalues with zero real part. This appendix is meant to summarize those aspects of the theory relevant for this text. All of the results are taken from Chapter 2 of Carr (1981), though the interested reader is encouraged to see Wiggins (1990) or Guckenheimer and Holmes (1983) as well.

By a suitable choice of basis, the dynamic about an ESS for the models of this text can be put in the form

$$\dot{x} = f(x, \; y)$$
$$\dot{y} = By + g(x, \; y) \tag{2.10.1}$$

where $x \in R^l$ and $y \in R^m$; B is an $m \times m$ matrix all of whose eigenvalues have negative real part; f and g are smooth functions that vanish together with their derivatives at the origin. Thus, the only eigenvalue with nonnegative real part in our applications is zero and the corresponding eigenspace has dimension l.

Furthermore, for our purposes, the variables x and y are often restricted in that certain of their components such as those of x are always nonnegative as in Property 5, Section 2.6 of Carr (1981). The following four theorems are those relevant for this text.

THEOREM 2.10.1. The system (2.10.1) has a local invariant centre manifold of the form $y = h(x)$ where h is a smooth function that vanishes together with its derivatives at the origin. The flow on this centre manifold is of the form

$$\dot{u} = f(u,\, h(u)) \qquad\qquad (2.10.2)$$

THEOREM 2.10.2. The zero solution of (2.10.1) with respect to the restricted dynamic is neutrally stable (respectively, l.a.s.) if and only if the zero solution of (2.10.2) is neutrally stable (respectively, l.a.s.).

THEOREM 2.10.3. The centre manifold contains any submanifold of equilibrium points that passes through the origin (Property 3, Section 2.6 of Carr).

THEOREM 2.10.4. If the zero solution of (2.10.2) is neutrally stable, then any solution of (2.10.1) that is initially sufficient close to the origin approaches asymptotically a solution of (2.10.2) that begins on the centre manifold.

3. FREQUENCY-DEPENDENT EVOLUTION IN A TWO-SPECIES HAPLOID SYSTEM

Chapter 2 emphasized the dynamic consequences of the original ESS theory that Maynard Smith (1974; 1982) developed to explain the observed frequency evolution of behavioral types in a single species. At the same time, the presentation provided a foundation from which to generalize evolutionary game theory to other biological systems. This and the following two chapters examine extensions of the original theory. By the end of Chapter 5, I expect all readers to appreciate both the power of ESS theory in these particular biological settings and the enormous potential of game-theoretic reasoning in general evolutionary processes.

A quick glance at the Table of Contents for Chapters 2 and 3 shows an obvious parallel between the theoretical development of their respective biological models. That is, Chapter 3 follows the four-step ESS program, proposed in Section 2.6, that generates ESS stability conditions in new biological systems. I consider the recent development (Cressman and Dash 1991) of evolutionary game theory for two-species frequency-dependent models to be of utmost importance. Indeed, for me, the material in Chapter 3 answers the question of why a text explaining the parallels between diverse theoretical evolutionary processes is necessary at this time.

3.1 Frequency-Dependent Fitness

Let I and J denote two species and suppose individuals of species I are characterized by (possible mixed) phenotypes $S \in \Delta^m$ and those of species J by $T \in \Delta^n$. Two-species frequency evolution will be based on individual fitnesses that depend only on the individual's phenotype, the strategy frequency of its conspecifics and the strategy frequency of the other species. In particular, in this chapter, fitness is independent of the relative densities of the two species (i.e. of the total number of individuals of one species relative to the other). This is a serious restriction; the consequences of which are not fully understood (Chapter 5 will examine density effects in a single species). On the other hand, the frequency-dependent coevolution examined in this chapter applies to any two-species ecosystem displaying stable co-existence at fixed relative density. It also applies in situations where there are no intraspecific interactions. For example, males and females may be represented as separate species to model certain male-female conflicts such as the one explored in Section 3.3 (A) below.

As in Chapter 2, the connection with matrix games is most natural when fitness changes result from pairwise contests. Suppose, on average, an individual in species I engages in intraspecific versus interspecific contests in the ratio c^I versus $1 - c^I$. [If there are no intraspecific interactions, then $c^I = 0$.] Then an individual $S_1 \in \Delta^m$ will have an expected fitness change in a random contest of $S_1 \cdot (A\mu + B\upsilon)$ where A and B are the $m \times m$ and $m \times n$ payoff matrices multiplied by the constants c^I and $1 - c^I$ respectively. Here μ and υ are the mean strategies of species I and J respectively. Similarly, a T_1 individual in J has fitness change $T_1 \cdot (C\mu + D\upsilon)$ per random contest. [Here C is an $n \times m$ and D is an $n \times n$ matrix. The multiplication factors, c^I and $1 - c^I$ as well as c^J and $1 - c^J$, are a mathematical nuisance. Henceforth, they will be included in the four payoff matrices which will still be labelled A, B, C, and D.]

3.2. Monomorphic ESS's and Stability

Suppose all individuals of species I initially present have phenotype S^* and those of species J have T^*. A two-species system at this monomorphic pair (S^*, T^*) is said to be evolutionarily stable if it cannot be invaded by a small (relative to the number in the original system) subsystem specified by the monomorphic pair (S, T).

Let ε and δ be the small proportions of S and T individuals in species I and J respectively. Then $\mu = \varepsilon S + (1 - \varepsilon)S^*$ and $\upsilon = \delta T + (1 - \delta)T^*$ represent the respective mean strategies of the two species. As in Chapter 2, we seek a heuristic involving comparisons of the fitness of S to S^* and of T to T^* in this perturbed system that will guarantee the mean strategies return to S^* and T^*. If S and T both do as well as S^* and T^* respectively, it seems intuitively clear that the system would evolve away from (S^*, T^*) towards (S, T) or, at best, (S^*, T^*) would exhibit a type of neutral stability.

What is surprising is that the converse is also true for the haploid dynamic. That is, (S^*, T^*) is noninvadable by the monomorphic pair (S, T) (see Theorem 3.2.4 below) if either S^* or T^* does better than S or T respectively in the system that is slightly perturbed from (S^*, T^*). [From the logic surrounding inequality (2.2.1), we might have expected to need both S^* and T^* do as well as S and T respectively and that at least one should do better.] Thus, in hindsight, we have

DEFINITION 3.2.1. (S^*, T^*) is a *monomorphic ESS* if, for all other monomorphic pairs (S, T) and for all positive ε and δ sufficiently small, at least one of the inequalities

$$S \cdot (A\mu + B\upsilon) < S^* \cdot (A\mu + B\upsilon) \quad \text{or} \tag{3.2.1}$$

$$T \cdot (C\mu + D\upsilon) < T^* \cdot (C\mu + D\upsilon) \tag{3.2.2}$$

is true where $\mu = \varepsilon S + (1 - \varepsilon)S^*$ and $\upsilon = \delta T + (1 - \delta)T^*$.

Although Definition 3.2.1 is the best heuristic formulation for monomorphic ESS's (it is the formulation emphasized for most of this chapter), the inequalities are surprisingly difficult to verify mathematically for a given choice of four payoff matrices. In fact, the following equivalent definition is more useful when finding ESS's in practical applications of the theory.

DEFINITION 3.2.2. (S^*, T^*) is a *monomorphic ESS* if, for all other monomorphic pairs (S, T), the following five conditions are satisfied.

(i) $S \cdot (AS^* + BT^*) \leq S^* \cdot (AS^* + BT^*)$

(ii) $T \cdot (CS^* + DT^*) \leq T^* \cdot (CS^* + DT^*)$

(iii) $(S - S^*) \cdot A(S - S^*) < 0$ if $S \neq S^*$ and there is equality in (i)

(iv) $(T - T^*) \cdot D(T - T^*) < 0$ if $T \neq T^*$ and there is equality in (ii)

(v) Either $(S - S^*) \cdot B(T - T^*) \leq 0$ and $(T - T^*) \cdot C(S - S^*) \leq 0$ or
$$[(S - S^*) \cdot A(S - S^*)][(T - T^*) \cdot D(T - T^*)]$$
$$> [(S - S^*) \cdot B(T - T^*)][(T - T^*) \cdot C(S - S^*)] \text{ if } S \neq S^*, T \neq T^* \text{ and there}$$
is equality in (i) and (ii).

In both definitions, the static ESS conditions are complicated by the arbitrary choice of possible individual strategies exhibited by the two species. As we will see in Section 3.4, these complications become more manageable when generalized to polymorphic populations where individuals may use any strategy in the species appropriate simplex (i.e. Δ^m or Δ^n). In the meantime, we develop the theory of monomorphic ESS's by first proving

THEOREM 3.2.3. Definitions 3.2.1 and 3.2.2 are equivalent.

PROOF. Expansions of (3.2.1) and (3.2.2) yield the following equivalent inequalities

$$S \cdot (AS^* + BT^*) + \varepsilon x \cdot Ax + \delta x \cdot By < S^* \cdot (AS^* + BT^*) \qquad (3.2.3)$$

$$T \cdot (CS^* + DT^*) + \varepsilon y \cdot Cx + \delta y \cdot Dy < T^* \cdot (CS^* + DT^*) \qquad (3.2.4)$$

where $x = S - S^* \in X^m$ and $y = T - T^* \in X^n$ (X^m and X^n are as in Definition 2.4.3).

Suppose (S^*, T^*) satisfies the five monomorphic ESS conditions of Definition 3.2.2. If (i) is a strict inequality, then (3.2.3) follows whenever ε and δ are sufficiently close to zero. Similarly, strict inequality in (ii) implies (3.2.4). The only situation left to consider is equality in both (i) and (ii). Under these equalities, if $x \cdot By \leq 0$ and $y \cdot Cx \leq 0$, both (3.2.3) and (3.2.4) are true by (iii) and (iv). On the other hand, if $x \cdot By > 0$, then (v) implies $y \cdot Cx < (x \cdot Ax y \cdot Dy)/(x \cdot By)$. In this case, for all positive ε and δ,

$$\varepsilon y \cdot Cx + \delta y \cdot Dy > \left(\varepsilon \frac{x \cdot Ax}{x \cdot By} + \delta\right) y \cdot Dy.$$

Since $y \cdot Dy < 0$, either $\varepsilon y \cdot Cx + \delta y \cdot Dy < 0$ or $\varepsilon x \cdot Ax + \delta x \cdot By < 0$. That is, (S^*, T^*) satisfies Definition 3.2.1. [A similar argument will also show that either (3.2.3) or (3.2.4) is true when $y \cdot Cx > 0$.]

Conversely, assume that either (3.2.1) or (3.2.2) holds. If (S^*, T^*) and (S, T) are monomorphic pairs, then so are (S^*, T) and (S, T^*). Suppose (S, T^*) is different than (S^*, T^*). Since (3.2.2) cannot be true for (S, T^*), (3.2.1) must be by the above assumption. From (3.2.3),

$$S \cdot (AS^* + BT^*) + \varepsilon x \cdot Ax < S^* \cdot (AS^* + BT^*)$$

for all positive ε sufficiently close to zero. Conditions (i) and (iii) now follow from the methods for a single species used in Section 2.2. Similarly, (ii) and (iv) result by consideration of the pair (S^*, T). To complete the proof, we must show (v) is true whenever (i) and (ii) are equalities. If $x \cdot By \leq 0$ and $y \cdot Cx \leq 0$ we are done. Without loss of generality, take $x \cdot By > 0$. Choose positive ε and δ sufficiently small with $\varepsilon x \cdot Ax + \delta x \cdot By = 0$. From (3.2.4), $\varepsilon y \cdot Cx + \delta y \cdot Dy < 0$. Thus

$$0 < \varepsilon x \cdot Ax(\varepsilon y \cdot Cx + \delta y \cdot Dy)$$
$$= \varepsilon x \cdot Ax \, \delta y \cdot Dy - \delta x \cdot By \, \varepsilon y \cdot Cx$$
$$= \varepsilon \delta (x \cdot Ax \, y \cdot Dy - x \cdot By \, y \cdot Cx)$$

That is, the strict inequality in (v) holds.

∎

Although the above result is impressive and will be most useful, I must caution the reader *at this point* that we have not justified either monomorphic ESS definition on dynamical grounds. The rest of the section characterizes the dynamic stability of monomorphic ESS's.

Suppose that the monomorphic pair (S, T) invades a system currently at a different monomorphism (S^*, T^*). As in Chapter 2, it is the continuous dynamic involving haploid species that most clearly establishes the equivalence (Theorem 3.2.4) of the monomorphic ESS conditions and dynamic stability.

Let N_1^I and N_2^I (respectively, N_1^J and N_2^J) represent the number of individuals using strategies S^* and S (respectively, T^* and T). Since the payoff matrices A, B, C and D already reflect the ratio of intraspecific versus interspecific contests, the continuous dynamic analogous to (2.2.5) is

$$\dot{N}_1^I = N_1^I S^* \cdot (A\mu + B\upsilon)$$
$$\dot{N}_2^I = N_2^I S \cdot (A\mu + B\upsilon)$$

(3.2.5)

$$\dot{N}_1^J = N_1^J T^* \cdot (C\mu + D\upsilon)$$
$$\dot{N}_2^J = N_2^J T \cdot (C\mu + D\upsilon)$$

(3.2.6)

where we have $\mu = (N_1^I S^* + N_2^I S)/N^I$, $\upsilon = (N_1^J T^* + N_2^J T)/N^J$, $N^I = N_1^I + N_2^I$, and $N^J = N_1^J + N_2^J$. If the problem of extinction (of either species) is again ignored as it was in the proof of Theorem 2.2.2, the frequency dynamic is given by the two-dimensional system

$$\dot{p} = p(S - \mu) \cdot (A\mu + B\upsilon)$$
$$\dot{q} = q(T - \upsilon) \cdot (C\mu + D\upsilon) \qquad (3.2.7)$$

were p is the proportion N_2^I/N^I of S-users in species I and q is the proportion of T-users. Thus, $\mu = pS + (1 - p)S^*$ and $\upsilon = qT + (1 - q)T^*$. [The reader should compare this with the dynamic (2.2.7).]

The justification of Definitions 3.2.1 and 3.2.2 is

THEOREM 3.2.4. (S^*, T^*) is a monomorphic ESS if and only if (S^*, T^*) is l.a.s. under the dynamic (3.2.7) for any other monomorphic pair (S, T).

PROOF. First consider an invasion of (S^*, T^*) by a monomorphism of the form (S, T^*) where S is the strategy of an individual in species I and $S \neq S^*$. Then $\upsilon = T^*$, $\dot{q} = 0$ and the dynamic (3.2.7) reduces to the single differential equation
$$\dot{p} = p(S - \mu) \cdot (A\mu + BT^*).$$
By Theorem 2.2.2 for a single species, $\dot{p} < 0$ for all positive p sufficiently close to zero iff conditions (i) and (iii) of Definition 3.2.2 are satisfied. Similarly, (ii) and (iv) are equivalent to l.a.s. under invasion by (S^*, T).

Thus the first four conditions of Definition 3.2.2 are necessary for l.a.s. and, conversely, they imply l.a.s. when invaded by a rare mutant in exactly one of the species. The proof is completed by assuming $S \neq S^*$ and $T \neq T^*$ and then showing $(p, q) = (0, 0)$ is l.a.s. under (3.2.7) iff (S^*, T^*) is an ESS. To see this rewrite the dynamic in terms of $x = S - S^*$ and $y = T - T^*$ as
$$\dot{p} = p(1 - p)x \cdot (AS^* + BT^* + pAx + qBy).$$
$$\dot{q} = q(1 - q)y \cdot (CS^* + DT^* + pCx + qDy). \qquad (3.2.8)$$

and divide the proof into parts.

(Necessity) Assume (S^*, T^*) is an ESS and consider the following three cases.
Case *(a)*: (Both (i) and (ii) are inequalities). Then p and q will both decrease monotonically to zero if p and q are initially positive and sufficiently close to zero.
Case *(b)*: (Exactly one of (i) and (ii), say (i), is an inequality). If $y \cdot Cx \leq 0$, then q will decrease monotonically to zero and so p also decreases to zero when q is initially close to zero.

On the other hand, if $y \cdot Cx > 0$, then $\dot{q} = 0$ along the line $py \cdot Cx + qy \cdot Dy = 0$ through the origin with positive slope depicted in Figure 3.2. Since (i) is an inequality, there is a rectangle with this line as diagonal as indicated in Figure 3.2 inside which $\dot{p} < 0$. Since $\dot{q} < 0$ in the top half of this rectangle, the dynamic follows the vector field as drawn and ensures (p, q) approaches $(0, 0)$ if initially inside this rectangle.

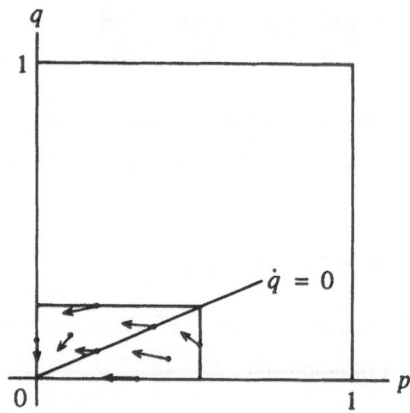

Figure 3.2. *The phase portrait of (3.2.8) in case (b) of Theorem 3.2.4*

Case (c):   (Equality in (i) and (ii)). This last case is the most interesting, especially when $x \cdot By > 0$ and $y \cdot Cx > 0$. If either of these inner products is nonpositive, then either p or q decreases monotonically to zero.

Thus, assume $x \cdot By$ and $y \cdot Cx$ are both positive and let r be the positive number satisfying $x \cdot By = ry \cdot Cx$. Then $(1 - p)(1 - q)^r$ is a Liapunov function since

$$\frac{d}{dt}\ln(1 - p)(1 - q)^r = -\left(\frac{\dot{p}}{1 - p} + r\frac{\dot{q}}{1 - q}\right)$$

$$= -(p^2 x \cdot Ax + pq(x \cdot By + ry \cdot Cx) + q^2 ry \cdot Dy)$$

is positive for all p and q between 0 and 1. This positivity, pointed out by Schuster et al. (1981), follows from the negative-definiteness of the 2×2 symmetric matrix

$$\begin{bmatrix} x \cdot Ax & x \cdot By \\ ry \cdot Cx & ry \cdot Dy \end{bmatrix}.$$

An added bonus in this case is that $(0, 0)$ is in fact globally stable.

Thus, a monomorpic ESS is l.a.s. The discussion in the first paragraph of the proof shows l.a.s. implies the first four conditions of Definition 3.2.2. The following argument completes the proof.

(Sufficiency) Assume $(0, 0)$ is l.a.s. under (3.2.8). To show condition (v) of Definition 3.2.2, note that it is certainly true if either $x \cdot By \leq 0$ or $y \cdot Cx \leq 0$. On the other hand, if both these inner products are positive and (v) were not true, then

$$x \cdot Ax \, y \cdot Dy \leq x \cdot By \, y \cdot Cx. \tag{3.2.9}$$

In this case, choose an initial p and q so that $\dot{p} = p(1 - p)(px \cdot Ax + qx \cdot By) = 0$.

From (3.2.9), $\dot{q} \geq 0$ and this in turn implies $\dot{p} \geq 0$ for all positive t. This monotonic increasing of p contradicts the l.a.s. of $(0, 0)$ and so (v) must be true.

∎

3.3 Examples: Battle-of-the-Sexes and Edgeworth Market Games

Each of the two examples developed in this section illustrates one of the monomorphic ESS definitions given in Section 3.2. At the same time, these examples are important to understand heuristically the discussions of the (polymorphic) ESS's in Section 3.4 and the polymorphic dynamic in Section 3.5.

The Battle-of-the-Sexes example (Maynard Smith 1982; Hofbauer and Sigmund 1988) is a common biological application of evolutionary games based on the two-species frequency dynamic of Section 3.2. This dynamic models the qualitative discussion of Dawkins (1976) concerning the male-female conflict over parental care of offspring. [A different example with the same name is often used in standard game-theory texts (Luce and Raiffa 1957; Owen 1982) to illustrate possible shortcomings of non-cooperative game theory applied to male-female interactions. If you are familiar with this literature, please disregard it here.]

The second example, Edgeworth Market Games, is nonconventional in the context of evolutionary game theory. Rather, it is a standard example taken directly from economic game theory (Owen 1982; Shubik 1959) that illustrates coalition formation and cooperation between commodity traders. I have included it here to convince biologists (and game theorists) that individual selection principles, on which evolutionary game theory is based, lead naturally to cooperative behavior (an unexpected inference since cooperation is intuitively associated with group selection arguments). In particular, this example connects the solution concepts of cooperative game theory with the heuristic ESS conditions of Definition 3.2.1.

(A) Battle-of-the-Sexes

Suppose adult males and females each exhibit two pure strategies in their mating and parental care behaviors; namely, females always care for their offspring but they can either be coy (insist on a long courtship before mating) or fast (mate immediately) while males are either faithful (willing to engage in long courtship and also care for the young) or philandering (will not wait nor care for their young). These behaviors, named by Dawkins (1976), will obey the formalism developed in Sections 3.1 and 3.2 under the following assumptions.

(i)   Individual fitness is completely determined by these behaviors.

(ii)   Each type of mating pair produces, on average, offspring with 1:1 sex ratio.

(iii) Offspring inherit the identical strategy as their parent of the same sex.

Let females be type I individuals and males type J. These may be regarded as different species with only interspecific (i.e. female versus male) interactions having fitness consequences. That is, A and D are the zero matrices and the dynamic, such as (3.2.7), only involves the 2×2 matrices B and C.

Intuition would suggest no monomorphic ESS can exist if all individuals must use pure strategies. For instance, a population of faithful males and coy females could be successfully invaded by a population of fast females (and faithful males) since such females would have the same reproductive success while avoiding the cost of courtship. The fast females and faithful males could, in turn, be invaded by fast females and philandering males, etc. Indeed, under the reasonable ordering of the payoffs associated to different male-female pairs considered in this section, the above intuition is correct (other orderings are considered in Thomas (1984)). At the same time as these qualitative aspects are shown, a more thorough analysis of mixed-strategy effects is explained than that done by other researchers.

For a quantitative analysis, suppose the cost and benefits of parental care from highest to lowest are as follows (the numbers in brackets give the payoffs considered by Dawkins (1976)).

(i)   the total cost of raising an offspring (20)
(ii)   the individual fitness gain of producing an offspring (15)
(iii) the individual cost of courtship (3)
(iv)   the fitness of not mating (0)

Thus, for example, the fitness of each parent in a coy-faithful mating is the gain (15) minus the courtship cost (3) and half of the cost of parental care (10). Hofbauer and Sigmund (1988)

Table 3.3. *The bimatrix payoffs for the Battle-of-the Sexes*

The entry above the diagonal in each of the four pure-strategy confrontations gives the payoff to the male while the entry below is the female's payoff.

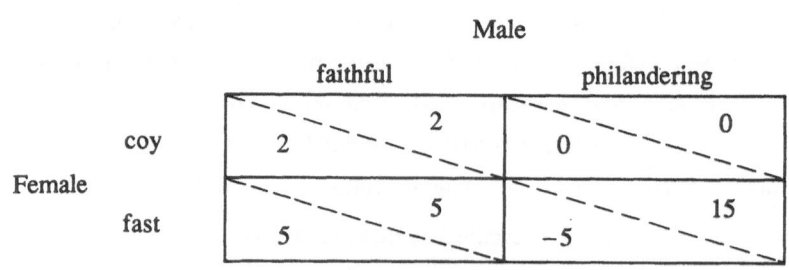

have shown no monomorphism is an ESS against invasion by one of the four possible pure-strategy pairs if the coy-faithful pair is more fit than when no mating occurs (i.e. 15-3-10 > 0).

Since there are no intraspecific interactions (i.e. no interactions between individuals of the same sex), the only relevant payoff matrices, B and C, can both be included in the *bimatrix* of Table 3.3 that uses the format of Maynard Smith (1982). That is, $B = \begin{bmatrix} 2 & 0 \\ 5 & -5 \end{bmatrix}$ and

$C = \begin{bmatrix} 2 & 5 \\ 0 & 15 \end{bmatrix}$. For the hawk-dove game of Section 2.3, when the cost of fighting is higher than the value of the resource, the mixed-strategy ESS S_0^* satisfying $AS_0^* = \lambda(1, 1)$ played a central role. For the current example, it will turn out that the mixed-strategy pair (S_0^*, T_0^*), given by $S_0^* = (5/6, 1/6)$ and $T_0^* = (5/8, 3/8)$, is just as important in determining monomorphic ESS's since

$$CS_0^* = 5/2 (1, 1) \text{ and } BT_0^* = 5/4 (1, 1). \tag{3.3.1}$$

Suppose (S^*, T^*) is a monomorphic ESS. The absence of A and D in Definition 3.2.2 implies (S^*, T^*) satisfies

$$\begin{aligned} S \cdot BT^* &< S^* \cdot BT^* \text{ if } S \neq S^* \\ T \cdot CS^* &< T^* \cdot CS^* \text{ if } T \neq T^* \end{aligned} \tag{3.3.2}$$

for any other monomorphic pair (S, T). By (3.3.1) and (3.3.2), $S^* \neq S_0^*$, $T^* \neq T_0^*$ and, furthermore, S_0^* and T_0^* are not both equal to convex combinations of S^*, S and T^*, T

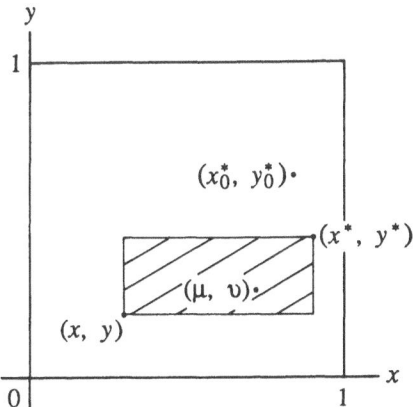

Figure 3.3.1. *Monomorphic ESS's for Battle-of-the Sexes*

Strategy-pairs are represented as points in the unit square. The equilibrium point (x_0^*, y_0^*) is $(5/6, 5/8)$. If all individual strategy-pairs (x, y) lie in the shaded rectangle, then so does the mean strategy (μ, υ). Furthermore, (x^*, y^*) is the unique monomorphic ESS in the case illustrated.

respectively. These restrictions can be understood best by representing the strategy pair (S, T) geometrically in Figure 3.3.1 as the point (x, y) in the unit square $[0, 1] \times [0, 1]$ where $S = (x, 1 - x)$ and $T = (y, 1 - y)$. The above asserts that (S^*, T^*), represented by (x^*, y^*), is not on the vertical or horizontal line through $(x_0^*, y_0^*) = (5/6, 5/8)$ and that the rectangle generated by the diagonal (x^*, y^*) to (x, y) does not include (x_0^*, y_0^*). In fact, the complete characterization is

THEOREM 3.3.1. (x^*, y^*) is a monomorphic ESS for the Battle-of-the-Sexes game if and only if (x_0^*, y_0^*) is outside the rectangle formed by all possible individual strategy pairs (x, y) and, when rays (i.e. line segments) are drawn from (x_0^*, y_0^*) to points in this rectangle, (x^*, y^*) is the unique vertex that is furthest counterclockwise.

PROOF. Consider the inequalities (3.3.2) only for the case depicted in Figure 3.3.1; namely, $x \leq x_0^* < x^*$ and $y \leq y^* < y_0^*$. [The other seven cases are left to the reader.] For the matrices in Table 3.3,

$$(S - S^*) \cdot BT^* = (S - S^*) \cdot B(T^* - T_0^*)$$
$$= -8(x^* - x)(y_0^* - y^*)$$
$$(T - T^*) \cdot CS^* = (T - T^*) \cdot C(S^* - S_0^*)$$
$$= -12(x^* - x_0^*)(y^* - y_0^*).$$

These will both be negative iff (x^*, y^*) is the indicated vertex. ∎

A comparison with Section 2.3 shows again that the existence and description of the monomorphic ESS's depend critically on the set of initial individual strategy-pairs. In fact, dynamic consequences are even more interesting. No pure-strategy population can resist invasion; rather the polymorphic pure-strategy dynamic cycles counterclockwise around the equilibrium (x_0^*, y_0^*) with

$$V(p, q) = p^2(1 - p)^{10} q^3(1 - q)^5$$

a constant of motion for (3.2.7) where p is the frequency of fast females and q of philandering males. That is, (x_0^*, y_0^*) is a neutrally stable equilibrium unlike the l.a.s. mixed-strategy ESS of the hawk-dove game.

The mixed-strategy dynamic is somewhat more complicated. The flow of the mean strategy (μ, υ) maintains a counterclockwise tendency. If (x_0^*, y_0^*) is in the rectangle formed by the set of possible mean strategies there is cycling; otherwise, (μ, υ) approaches a point on the furthest counterclockwise ray. If this point is unique, it is a globally stable monomorphic ESS; otherwise it is only part of the neutrally stable line segment that forms an ES Set as discussed in Chapter 6.

(B) Edgeworth Market Games

Edgeworth (1881) introduced much of the modern game-theory approach to market games in his model of trading two commodities, A and B, between two types of players, I and J. [The terminology adopted here follows Shubik (1984) most closely.] A *commodity bundle* possessed by an individual is represented by the ordered pair (x, y) where x is the amount owned of commodity A and y is the amount of B. Intuitively, an individual will trade with another to obtain a preferred commodity bundle. To formalize this, we attach to each individual i a *preference relationship* $\geq_i$ that is a partial order on the possible commodity bundles satisfying the following four properties.

(i)   The order is complete. Given bundles (x_1, y_1) and (x_2, y_2), either
        $(x_1, y_1) \geq_i (x_2, y_2)$ or $(x_2, y_2) \geq_i (x_1, y_1)$ or both. If $(x_1, y_1) \geq_i (x_2, y_2)$
        and $(x_2, y_2) \geq_i (x_1, y_1)$, individual i is *indifferent* to the two bundles.

(ii)   Preference is transitive. If $(x_1, y_1) \geq_i (x_2, y_2)$ and $(x_2, y_2) \geq_i (x_3, y_3)$ then
        $(x_1, y_1) \geq_i (x_3, y_3)$.

(iii)   Preference increases in each commodity. $(x_1, y) \geq_i (x_2, y)$ if and only if
        $x_1 \geq x_2$. $(x, y_1) \geq_i (x, y_2)$ if and only if $y_1 \geq y_2$.

(iv)   Indifference curves are convex. For a given bundle (x_1, y_1), $\{(x, y) \mid$ individual i is
        indifferent to (x, y) and $(x_1, y_1)\}$ is a smooth convex curve (in the first quadrant).

If $(x_1, y_1) \geq_i (x_2, y_2)$ is translated as "individual i prefers bundle (x_1, y_1) to (x_2, y_2)", the properties seem reasonable and self-explanatory except perhaps the last. For us, it is sufficient to interpret (iv) as asserting that an individual, indifferent to the two bundles (x_1, y_1) and (x_2, y_2), prefers the average bundle $((x_1 + x_2)/2, (y_1 + y_2)/2)$ to both. Mathematically, the above properties are equivalent to the existence of a smooth real-valued *utility function* $u_i(x, y)$, defined for all bundles (x, y), whose convex level curves increase in level as either commodity increases. Such a utility function quantifies preference by letting $(x_1, y_1) \geq_i (x_2, y_2)$ if and only if $u_i(x_1, y_1) \geq u_i(x_2, y_2)$. [For instance, the smooth indifference curves in Figure 3.3.2 for individual I are sketches of the level curves of $u(x, y) = (1 + x)(1 + y)$.]

With these formalities in mind, let us return to the market game. Assume, as Edgeworth did, that all individuals of type I have the same preference relation as do all those of type J and that all type I individuals have an initial bundle $(a, 0)$ and J have $(0, b)$. It turns out that the solution to the simplest game considered by Edgeworth, the $[1, 1]$ game that has one individual of each type, is closely connected to the characterization of a monomorphic ESS in Definition 3.2.1.

For the [1, 1] game, let u_I and u_J denote the utility function of the individual of type I and J respectively. These two individuals trade commodities until a final *allocation* (i.e. a pair of bundles) is agreed upon. An effective way to compare Edgeworth's discussion to Definition 3.2.1, is to denote allocations as (S, T) where S and T are the commodity bundles of I and J respectively and are both points in the rectangle $\{(x, y)|0 \leq x \leq a, \ 0 \leq y \leq b\}$. The final allocation is a solution according to Edgeworth if it is preferred by both to their initial allocation and if no other feasible allocation is preferred by both to it. That is,

DEFINITION 3.3.2. (S^*, T^*) is a *solution* to the [1, 1] game if

(i) $S^* = (x^*, y^*)$ and $T^* = (a - x^*, b - y^*)$ have nonnegative components

(ii) $u_I(S^*) \geq u_I((a, 0))$ and $u_J(T^*) \geq u_J((0, b))$

(iii)   for any other allocation (S, T), either

$$u_I(S^*) > u_I(S) \text{ or } u_J(T^*) > u_J(T). \tag{3.3.3}$$

The first condition illustrates a critical difference, used to great advantage by Edgeworth, between his market games and the evolutionary games discussed so far in Chapter 3. Let us agree to call S and T strategies and so the allocation (S, T) is a strategy-pair. [Game theorists may cringe at this usage for market games.] Then strategy S for I depends on that of T for J, and vice versa. Specifically, if I uses strategy $S = (x, y)$, then automatically J will use $T = (a - x, b - y)$ since preferences always increase in each commodity. This interdependence allows the geometric construction of the Edgeworth Box. In it, strategy-pairs are represented by points in the rectangle $\{(x, y)_I | 0 \leq x \leq a, 0, \leq y \leq b\}$ where $(x, y)_I$ is I's strategy. The origin for I, $(x, y)_I = (0, 0)$, is the lower left vertex in Figure 3.3.2 while that for J, $(x, y)_I = (a, b)$, is the upper right.

The last condition of Definition 3.3.2 is the most interesting. [Readers familiar with Pareto optimality in two-person cooperative games (Owen 1982) may be concerned both inequalities in (3.3.3) are strict. The technical difficulty of equality is avoided here by assuming the indifference curves are strictly convex.] First, it shows that a solution in the interior of the Edgeworth Box must be a point $(x, y)_I$ where the indifference curves of I and J, both convex to the origin of that particular individual, meet in common tangency. Otherwise, both I and J could shift their strategy-pair to a mutually preferred alternative. In Figure 3.3.2, the set of all such solutions form the *contract curve* (Edgeworth 1881; Shubik 1984) lying between the two indifference curves (for I and J respectively) that go through the initial strategy-pair $((a, 0)_I, (0, b)_J)$.

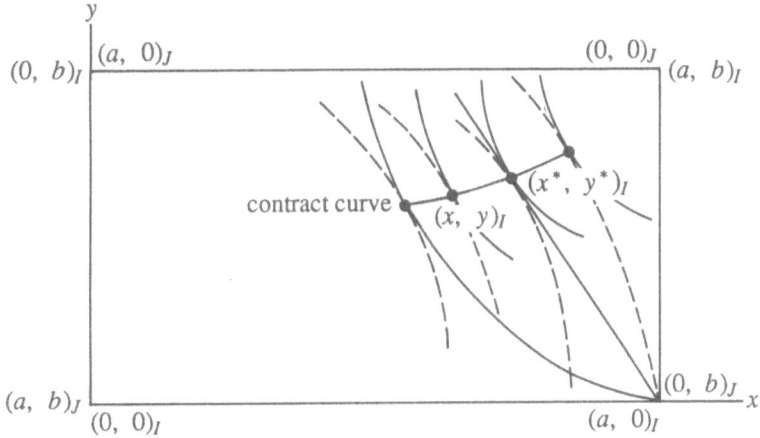

Figure 3.3.2. *The Edgeworth Box*

The convex indifference curves (solid for I and dashed for J) all have
negative slopes. The contract curve, that consists of points with common
tangent lines, goes through both $(x, y)_I$ and $(x^*, y^*)_I$, the latter being a
competitive equilibrium.

Secondly, for the [1, 1] game, the utility $u_I(S)$ of individual I (which we will also agree
is the fitness payoff) depends only on the strategy S and not on the system's mean strategy as in
Definition 3.2.1. Thus, the inequalities in (3.3.3) are the exact translation of the two fitness
comparisons (3.2.1) and (3.2.2). To reiterate this important conclusion for the [1, 1] game —
points on the contract curve (i.e. solutions) are the monomorphic ESS's of the two-species
system where utility functions are interpreted as individual fitness.

The correspondence between the dynamic consequences of Definitions 3.2.1 and 3.3.2 is not
as exact. Solutions on the contract curve cannot be l.a.s. like the monomorphic ESS's for the
matrix games of Section 3.2 since the utility functions are nonlinear. The location of the
neutrally stable final allocation on the contract curve depends on such intangibles as the relative
bargaining power of the two individuals.

A solution concept closer to l.a.s. is possible when the number of individuals of each type
grows arbitrarily large. The generalization will be summarized for the [n, n] market game
first from a static and then from a dynamic game-theoretic point of view. Then a biological
interpretation of these two methods will be discussed at the end of the section. Rigorous
definitions and results will be omitted to better concentrate on the relationship between ESS
principles and this solution concept.

Edgeworth (1881) showed that, as n becomes large, the $[n, n]$ market game with n individuals of each type has only isolated solutions, each corresponding to a monomorphic pair (S, T) where all individuals of the same type have the same final commodity bundle (i.e. strategy), and that (S, T) must still lie on the contract curve of Figure 3.3.2. [The assumption of an equal number of individuals of each type amounts to maintaining the fixed relative density used to justify the frequency approach in Section 3.1.] In fact, these solutions are the isolated points, such as $(x^*, y^*)_I$ in Figures 3.3.2 and 3.3.3, on the contract curve where the common tangent line passes through the initial allocation and are called *competitive equilibria*. In modern terms, Edgeworth's static approach is equivalent to finding the *core* of a $2n$-person cooperative game (Owen 1982; Shubik 1984). Specifically, at any other allocation, not in the core, some *coalition* (i.e. subset of the $2n$ individuals) can form and redistribute their resources among themselves in such a way that all coalition members do at least as well and some do better (i.e. have higher utility) than before redistribution.

An alternative dynamic game-theoretic approach to competitive equilibria had its beginnings before the work of Edgeworth (Chapters 11 and 12, Arrow and Hahn 1971). The approach assumes that the relative prices of the two commodities are set by an external agent (an *auctioneer*) and the n individuals of each type must declare quantities they are willing to trade at these prices. If the total supply and demand are unequal for one of the commodities, the auctioneer raises the relative price of the commodity in excess demand. Then individuals declare their revised trading intentions at the new prices and the process continues until equilibrium occurs.

A modification of the Edgeworth Box, shown in Figure 3.3.3, is instructive here as well. The auctioneer's prices are represented by lines with negative slopes through the initial allocation — steeper lines reflect higher relative prices of A. Line $Q_1 Q_2$ has the price of A equal to the price of B (since it has slope -1). At this price, I will trade two A's for two B's since I's indifference curve is tangent at Q_2 while J trades three B's for three A's. Thus the demand for A exceeds its supply and so the relative price of A should be raised towards that represented by the locally stable competitive equilibrium $(x^*, y^*)_I$.

The above dynamic is based on the actions of an auctioneer outside the control of the system. In biological terms, this external force can be regarded as a changing environment affecting the payoffs over time. However, it is possible to derive the properties of competitive equilibria from dynamic considerations of internally-controlled frequency-dependent and/or density-dependent payoffs — the usual vantage point of this text.

One way to see this is to represent the monomorphic pair $(x, y)_I$ in Figures 3.3.2 and 3.3.3 as a large (but equal) number of individuals of both type, each of whom is willing to exchange a

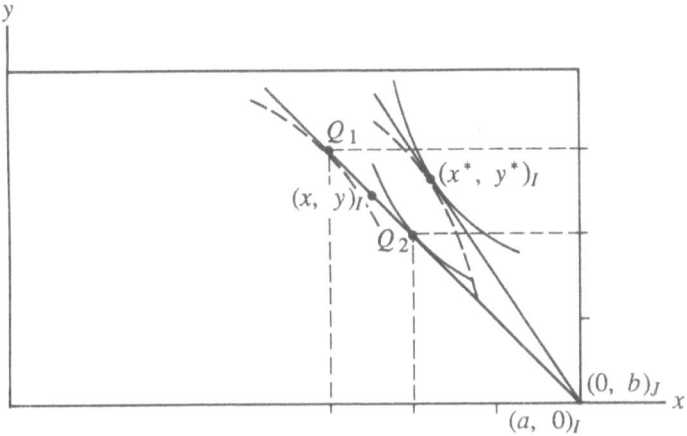

Figure 3.3.3. *A modified Edgeworth Box*

The two points, $(x, y)_I$ and $(x^*, y^*)_I$, on the contract curve are the same as those in Figure 3.3.2. At equal prices for the two commodities, I trades at Q_2 while J trades at Q_1. Therefore, $(x, y)_I$ is not a competitive equilibrium. As the auctioneer raises the price of A, both types of individuals tend towards trading at the competitive equilibrium $(x^*, y^*)_I$.

small number of A's and the same number of B's without marginally affecting their utilities. If the tangent line to I's indifference curve at $(x, y)_I$ intersects the x-axis before the initial allocation, then a mutant I with initial bundle $(a, 0)$ can invade by playing the field against (i.e. by trading with) the J individuals in the monomorphism and thereby obtain the higher utility available at $(x^*, y^*)_I$. Thus, any monomorphism that is not a competitive equilibrium can be successfully invaded. At the same time, density effects can also be included by allowing different numbers of type I individuals as opposed to type J. The Edgeworth Box that plots the average per capita allocation of each type is altered to plot the entire allocation. The sides of the box will change as densities change; however, the indifference curves remain the same. This has the effect of encouraging the appearance of more I individuals as opposed to J if commodity A is in excess demand. In this scenario, a combination of price and density changes continues as the system approaches competitive equilibrium.

3.4 The Static Characterization of a Two-Species ESS

The characterization of monomorphic ESS's for the Battle-of-the-Sexes example showed a complicated dependence on the initial set of individual strategy-pairs. As in Section 2.4, a polymorphic approach provides a better conceptual basis of the ESS conditions.

Suppose the mean strategies of the two species, I and J, are currently at S^* and T^*, respectively. If $(S^*, T^*) \in \Delta^m \times \Delta^n$ is to be evolutionarily stable, it must resist invasion by any subsystem whose pair of mean strategies is an arbitrary point in $\Delta^m \times \Delta^n$. That is, for the matrix games of Sections 3.1. and 3.2, the five conditions of Definition 3.2.2 must hold at an ESS (S^*, T^*) for any other $(S, T) \in \Delta^m \times \Delta^n$. [This is the analogue of Definition 2.4.1 and can be used much as in Theorems 2.4.4 and 2.4.5 to produce a short list of possible ESS's. For instance, it is not hard to show an ESS in the interior of $\Delta^m \times \Delta^n$ is unique.]

Alternatively, the following, more heuristic, (polymorphic) ESS definition is based on Definition 3.2.1. Analogous to Section 2.4, mean strategies for polymorphic species I and J will be denoted by S and T respectively (instead of μ and υ as in Section 3.2). The adjustment should be less painful this time around! This definition implements STEP 3 of Section 2.6. The crucial dynamical justification of STEP 4 must await Section 3.5.

DEFINITION 3.4.1. (S^*, T^*) is an *ESS* if, for all other (S, T) in some neighborhood of (S^*, T^*) in $\Delta^m \times \Delta^n$, either

$$S^* \cdot (AS + BT) > S \cdot (AS + BT) \quad \text{or} \qquad (3.4.1)$$

$$T^* \cdot (CS + DT) > T \cdot (CS + DT). \qquad (3.4.2)$$

Thus, if there are individuals playing S^* and T^*, then the ESS (S^*, T^*) will be a monomorphic ESS irrespective of the other monomorphic pairs under consideration. However, by Section 3.3 (A), the converse is not true. In particular, the Battle-of-the-Sexes game has no ESS. This can be seen most easily by first verifying no pure strategy-pair ESS exists and then applying the following result (Selten 1980) for general bimatrix (i.e. A and D are the zero matrices) games.

THEOREM 3.4.2. If (S^*, T^*) is an ESS of a bimatrix game, then S^* and T^* are pure strategies in Δ^m and Δ^n respectively.

PROOF. Suppose S^* has more than one pure strategy of species I in its support. Take $T = T^*$ and apply Definition 3.4.1. Since $T^* \cdot CS = T \cdot CS$, (3.4.1) must be true (i.e. $S^* \cdot BT^* > S \cdot BT^*$) for all $S \in \Delta^m$ except $S = S^*$. This is impossible by considering two choices of S that have S^* as their average strategy (and whose supports are contained in that of S^*).

■

For two-species models with intraspecific payoff matrices, the characterization of an ESS equivalent to Definition 3.4.1 was first proposed by Cressman and Dash (1991). This characterization is different from the original proposals (Taylor 1979; Schuster et al. 1981; Hofbauer and Sigmund 1988) that are equivalent to requiring

$$S^* \cdot (AS + BT) + T^* \cdot (CS + DT) > S \cdot (AS + BT) + T \cdot (CS + DT) \qquad (3.4.3)$$

for any other $(S, T) \in \Delta^m \times \Delta^n$ in some neighborhood of (S^*, T^*). [Clearly, (3.4.3) implies (S^*, T^*) is an ESS. However, there are ESS's that do not satisfy (3.4.3).] Each side of inequality (3.4.3) is the sum of two payoffs — one for each species. Thus (3.4.3) can be interpreted as saying (S^*, T^*) has a higher total payoff than that of the population mean (S, T). Such a condition for evolutionary stability suggests a means exists to share the payoffs equally between individuals of the two different species. This is unwarranted both intuitively (how can individual reproductive success be shared with another species) and analytically (dynamic stability is characterized by (3.4.1) and (3.4.2)).

It is my opinion that the general acceptance of the misleading condition (3.4.3) has had a profound effect in delaying the development of ESS theory in two-species models. One explanation for this acceptance may be the fact that the primary application to date of evolutionary games incorporating the frequency dynamic of Section 3.2 has been to biological situations where there are no intraspecific interactions. For these bimatrix games, (3.4.3) and Definition 3.4.1 are the same.

I predict a rapid diversification of areas where evolutionary game theory is applied once the stability principle contained in Definition 3.4.1 is more widely known. ESS theory is the natural setting to analyze two-species population models when each species has diverse individual characteristics because the theory is based on individual interactions rather than on group interactions implicit in the usual population models involving homogeneous species. An important step in this diversification would be to develop coevolutionary ESS models that also involve population size. This is partially accomplished by the density-dependent ESS theory of a single species considered in Chapter 5 — but much work is left to be done.

On the other hand, the discussion of Edgeworth Market Games already portends a bright future for collaboration, on frequency-dependent models, between theoretical biologists and game theorists. The only difference between the core solution concept of cooperative game theory and Definition 3.4.1 is that the core stability conditions need only be verified locally in evolutionary game theory. This complements the intuition that the rational decision process of cooperative game theory allows the population to shift radically to mutually beneficial remote strategy-pairs whereas evolution permits only a gradual change. Although the non-matrix payoffs prevalent in market games complicates a direct comparison between the ESS of Definition 3.4.1 and the core concept of cooperative games, the linearization techniques of Section 2.9 generalized to two-species systems can be used to analyze the local stability of market games.

The collaboration referred to above can be beneficial to both groups of researchers. For instance, game theorists should be encouraged to re-examine the relevance of assuming rational players in their theories of cooperation and coalition formation. Furthermore, internal interactions should also be considered when modelling large-scale games such as international conflicts or management/union relations. The "sides" in these games should not be regarded as two monolothic players but rather reflect pressures inside (i.e. intraspecific pressures) each organization as well. On the other hand, biologists have the potential to gain new insights into stability of coevolutionary systems. Many evolutionary games have no ESS and, in these cases at least, the numerous alternative solution concepts of cooperative game theory need investigation for their dynamic consequences.

3.5. Strong Stability for the Continuous Dynamic

This section considers the dynamic consequences of Definition 3.4.1 for polymorphic populations, especially the stability of the system's mean strategy-pair (S, T) near an ESS (S^*, T^*). Suppose $S_1, S_2, ..., S_M$ and $T_1, T_2, ..., T_N$ list the (mixed) strategies in Δ^m and Δ^n exhibited by individuals in species I and J respectively. Let p_i (and q_j) be the time-dependent frequency of individuals using strategy S_i (and T_j) in species I (and J). Then

$$S = S(p) = \sum_{i=1}^{M} p_i S_i \text{ and } T = T(q) = \sum_{j=1}^{N} q_j T_j \text{ are the mean strategies of species } I \text{ and } J$$

respectively.

The haploid frequency dynamic (see (2.5.1) and (3.2.7)) is

$$\dot{p}_i = p_i(S_i - S) \cdot (AS + BT)$$
$$\dot{q}_j = q_j(T_j - T) \cdot (CS + DT)$$

(3.5.1)

where $1 \leq i \leq M$ and $1 \leq j \leq N$. This dynamic on $\Delta^M \times \Delta^N$ induces an evolution of S and T. If (S^*, T^*) is to be an equilibrium of (3.5.1) then S^* and T^* must be convex combinations of $\{S_1, ..., S_M\}$ and $\{T_1, ..., T_N\}$ respectively, say $S(p^*) = S^*$ and $T(q^*) = T^*$. In such a case, (S^*, T^*) will be called l.a.s. (see Definition 2.5.1) under (3.5.1) if (S, T) evolves to (S^*, T^*) whenever (S, T) is initially sufficiently close to (S^*, T^*). Moreover, we define, as in Section 2.6

DEFINITION 3.5.1. (S^*, T^*) is *strongly stable* if, whenever $S^* \in \Delta^m$ is a convex combination of $S_1, ..., S_M$ and $T^* \in \Delta^n$ is a convex combination of $T_1, ..., T_N$, (S^*, T^*) is a l.a.s. equilibrium for the mean strategy evolution of (S, T) determined by (3.5.1).

By applying this definition to the monomorphic pairs of Section 3.2, we have that any strongly stable (S^*, T^*) is an ESS by Theorem 3.2.4. What about the converse? That is, does strong stability give the complete dynamic characterization of an ESS as in Theorem 2.6.2?

My conjecture is that strong stability is equivalent to the ESS conditions but I am unable to prove this is degenerate cases.

The partial result given next relies on the linerization of the dynamic (3.5.1) and assumes a regularity condition similar to that of Theorem 2.8.3. To remove this restriction seems to require a deeper analysis of the resultant centre manifold (Cressman and Dash 1991), perhaps by adapting the generalized Shahshahani metric of Hines (1991).

THEOREM 3.5.2. Suppose (p^*, q^*) is an equilibrium of (3.5.1) that corresponds to an ESS (S^*, T^*). If (p^*, q^*) is *regular* (i.e. if $S_i \cdot (AS^* + BT^*) < S^* \cdot (AS^* + BT^*)$ whenever $p_i^* = 0$ and $T_j \cdot (CS^* + DT^*) < T^* \cdot (CS^* + DT^*)$ whenever $q_j^* = 0$), then $(S(p), T(q))$ will evolve to (S^*, T^*) if (p, q) is initially sufficiently close to (p^*, q^*).

PROOF. To linearize (3.5.1), suppose $p_i = p_i^* + x_i$ and $q_j = q_j^* + y_j$. Since $\dot{p}_i = \dot{x}_i$ and $p_i^*(S_i - S^*) \cdot (AS^* + BT^*) = 0$, we have

$$\dot{x}_i = p_i^* \left[(S_i - S^*) \cdot (A \Sigma x_k S_k + B \Sigma y_l T_l) - \Sigma x_k S_k \cdot (AS^* + BT^*) \right.$$

$$+ x_i(S_i - S^*) \cdot (AS^* + BT^*) + o(|x| + |y|).$$

Similarly

$$\dot{y}_j = q_j^* \left[(T_j - T^*) \cdot (C \Sigma x_k S_k + D \Sigma y_l T_l) - \Sigma y_l T_l \cdot (CS^* + DT^*) \right.$$

$$+ y_j(T_j - T^*) \cdot (CS^* + DT^*) + o(|x| + |y|).$$

By regularity, the only nonzero entry in the rows of the above linearized dynamic that correspond to $p_i^* = 0$ or $q_j^* = 0$ are negative diagonal eigenvalues.

Therefore, these rows (and columns) may be deleted to show stability and we may assume all p_i^* and q_j^* are positive. In particular, there is no restriction on the signs of x_i and y_j — only that they be in a sufficiently small neighborhood of the origin. This implies, since (S^*, T^*) is an ESS, that $S_k \cdot (AS^* + BT^*) = S^* \cdot (AS^* + BT^*)$ for all S_k. From this and a similar calculation involving species J, we have

$$\Sigma x_k S_k \cdot (AS^* + BT^*) = (\Sigma x_k)S^* \cdot (AS^* + BT^*)$$
$$\Sigma y_l T_l \cdot (CS^* + DT^*) = (\Sigma y_l)T^* \cdot (CS^* + DT^*).$$

The linearized dynamic has the form $\begin{pmatrix} \dot{x} \\ \dot{y} \end{pmatrix} = L \begin{pmatrix} x \\ y \end{pmatrix}$ where L is the square matrix of order $M + N$ that has block form

$$L = \begin{bmatrix} L^{II} & L^{IJ} \\ L^{JI} & L^{JJ} \end{bmatrix} \tag{3.5.2}$$

where, for instance, L^{II} and L^{IJ} are the $M \times M$ and $M \times N$ matrices with entries

$L_{ik}^{II} = p_i^*(S_i - S^*) \cdot A(S_k - S^*)$ and $L_{il}^{IJ} = p_i^*(S_i - S^*) \cdot B(T_l - T^*)$ respectively.
[The entries of L^{JI} and L^{JJ} are given by similar expressions involving q_j^*.]

As in the proof of Lemma 2.8.2, introduce the Shahshahani inner product on $X^M \times Y^N =$
$\{(x, y) \in R^M \times R^N \mid \Sigma x_k = 0, \ \Sigma y_l = 0\}$ given by

$$\langle\langle (x, y), (\xi, \eta) \rangle\rangle = \Sigma \frac{x_i \xi_i}{p_i^*} + \Sigma \frac{y_j \eta_j}{q_j^*}. \tag{3.5.3}$$

By Lemma 3.5.3 below, either

$$\langle\langle (x, 0), L(x, y) \rangle\rangle < 0 \quad \text{or}$$
$$\langle\langle (0, y), L(x, y) \rangle\rangle < 0. \tag{3.5.4}$$

for all $(x, y) \in X^M \times Y^N$ unless $\Sigma x_i S_i = 0$ and $\Sigma y_j T_j = 0$.

Suppose, for the moment, that $\{S_1, \cdots, S_M\}$ and $\{T_1, \cdots, T_N\}$ are both linearly independent sets of vectors in Δ^m and Δ^n respectively. [For instance, this is true if all these individual strategies are pure.] Then, by Lemma 3.5.4 below, all relevant eigenvalues of L have negative real part and (p^*, q^*) is l.a.s.

In general (i.e. without linear independence), the zero eigenspace of L restricted to the invariant subspace $X^M \times Y^N$ consists of $\{(x, y) \mid \Sigma x_i S_i = 0 \text{ and } \Sigma y_j T_j = 0\}$. Since this subspace also corresponds to the set of equilibria of (3.5.1.) (e.g. $S(p) = S(p^*)$ iff $\Sigma x_i S_i = 0$), it is the centre manifold. The proof is completed by an application of centre manifold theory, specifically Theorems 2.10.3 and 2.10.4.
∎

LEMMA 3.5.3. If (S^*, T^*) is an ESS and L is given by (3.5.2), then L satisfies (3.5.4).

PROOF. $\langle\langle (x, 0), L(x, y) \rangle\rangle = \sum_{i=1}^{M} \frac{x_i(L(x, y))_i}{p_i^*}$ by (3.5.3)

$$= \Sigma \frac{1}{p_i^*} x_i p_i^* (S_i - S^*) \cdot [A\Sigma x_k(S_k - S^*) + B\Sigma y_l(T_l - T^*)]$$

$$= \Sigma x_i S_i \cdot (A\Sigma x_k S_k + B\Sigma y_l T_l) \text{ since } \Sigma x_k = \Sigma y_l = 0$$

$$= (S - S^*) \cdot (AS + BT)$$

where $S = S^* + \Sigma x_k S_k$ and $T = T^* + \Sigma y_l T_l$ are in Δ^m and Δ^n respectively when (x, y) is sufficiently close to the origin of $X^M \times Y^N$. Similarly,

$$\langle\langle (0, y), L(x, y) \rangle\rangle = (T - T^*) \cdot (CS + DT).$$

That is, (3.5.4) is an immediate consequence of Definition 3.4.1 since $(S, T) \neq (S^*, T^*)$ unless $\Sigma x_k S_k = 0$ and $\Sigma y_l T_l = 0$.
∎

LEMMA 3.5.4. If the matrix L given by (3.5.2) satisfies (3.5.4) for all nonzero vectors $(x, y) \in X^M \times Y^N$, then all eigenvalues of L restricted to $X^M \times Y^N$ have negative real part.

PROOF. Suppose $a + bi$ is an eigenvalue of L where $a, b \in R$ and i is the complex root of -1. If $b = 0$, there is a nonzero vector $w = (x, y) \in X^M \times Y^N$ such that $Lw = aw$. Then $\langle\langle(x, 0), L(x, y)\rangle\rangle = \langle\langle(x, 0), a(x, y)\rangle\rangle = a\langle x, x\rangle$ and $\langle\langle(0, y), L(x, y)\rangle\rangle = a\langle y, y\rangle$ where (by abuse of notation) $\langle x, \xi\rangle = \sum \dfrac{x_i \xi_i}{p_i^*}$ and $\langle y, \eta\rangle = \sum \dfrac{y_j \eta_j}{q_j^*}$. Thus, by (3.5.4), $a < 0$ (i.e. the eigenvalue is negative).

For the rest of the proof, assume $b \neq 0$. There are then two linearly independent vectors $w = (x, y)$ and $w' = (x', y')$ in $X^M \times Y^N$ such that

$$Lw = aw - bw' \quad \text{and} \quad Lw' = aw' + bw.$$

(i.e. $L(w + iw') = (a + ib)(w + iw')$). In fact, w can be taken as any nonzero vector in $\text{span}\{w, w'\} = \{\alpha w + \beta w' \mid \alpha, \beta \in R\}$ (for such w, take $w' = (aw - Lw)/b$). Since $\langle\langle(x, 0), L(x, y)\rangle\rangle = \langle\langle(x, 0), (ax - bx', ay - by')\rangle\rangle = a\langle x, x\rangle - b\langle x, x'\rangle$ and also $\langle\langle(0, y), L(x, y)\rangle\rangle = a\langle y, y\rangle - b(y, y')$, by (3.5.4), either

$$a\langle x, x\rangle - b(x, x') < 0 \quad \text{or} \quad a\langle y, y\rangle - b(y, y') < 0. \tag{3.5.5}$$

By a similar calculation with $\langle\langle(x', 0), L(x', y')\rangle\rangle$ and $\langle\langle(0, y'), L(x', y')\rangle\rangle$, either

$$a\langle x', x'\rangle + b(x, x') < 0 \quad \text{or} \quad a\langle y', y'\rangle + b(y', y) < 0. \tag{3.5.6}$$

To complete the proof, we will assume $a \geq 0$ and reach a contradiction. First, none of the inequalities in (3.5.5) or (3.5.6) is zero. For example, if $a\langle x, x\rangle - b\langle x, x'\rangle = 0$, then also $a\langle y, y\rangle - b\langle y, y'\rangle < 0$ by (3.5.5). In particular, we must have $b\langle y, y'\rangle > 0$ and so $a\langle y', y'\rangle + b\langle y', y\rangle > 0$. Then, by (3.5.6), $a\langle x', x'\rangle + b\langle x, x'\rangle < 0$ and this, in turn, implies that $a\langle x, x\rangle - b\langle x, x'\rangle$ is not zero but positive. Second, if $\langle\langle(x, 0), L(x, y)\rangle\rangle = a\langle x, x\rangle - b\langle x, x'\rangle < 0$, then we have $\langle\langle(\xi, 0), L(\xi, \eta)\rangle\rangle < 0$ for all $\omega = (\xi, \eta)$ in $\text{span}\{w, w'\}$ (or else this inner product will be zero for some nonzero ω). This is clearly impossible since w' is in the above span and $\langle\langle(x', 0), L(x', y')\rangle\rangle$ is positive. A similar contradiction arises by assuming any of the other three inequalities in (3.5.5) and (3.5.6). ∎

3.6 Multi-Species Frequency-Dependent Evolution

I feel confident the conjecture of the previous section, asserting the equivalence of strongly stable (S^*, T^*) with two-species frequency-dependent ESS's, will eventually be proven and thus successfully complete the last step of the ESS program of Section 2.6. At the same time, this section shows I am much less certain about a similar correspondence for multi-species (i.e. more than two) frequency-dependent evolution.

A quick comparison of Definitions 2.4.2 and 3.4.1 suggests that a three-species ESS (S^*, T^*, R^*) should have at least one of the strategies (e.g. S^*) more fit than that of the same species' mean strategy (i.e. S) when the system's mean strategies are slightly perturbed to (S, T, R). Let us refer to this idea as Attempt A at an ESS definition for three-species models. As the name implies, Attempt A does have some inadequacies. For instance, the following example proves stability for the pure-strategy model need not hold at such an (S^*, T^*, R^*).

To emphasize the essential mathematical point, the counterexample will assume each of the three species has exactly two pure strategies and that all three equilibrium strategies S^*, T^*, and R^* are equal to the vector $(1/2, 1/2) \in \Delta^2$. The frequency-dependent haploid dynamic is then based on a 6×6 payoff matrix of the form

$$\begin{bmatrix} A & B & C \\ D & E & F \\ G & H & I \end{bmatrix} \tag{3.6.1}$$

where the nine 2×2 submatrices represent payoffs for individual contests. [For instance, the entries in B give the interspecific payoffs of pure strategists in the first species when competing against pure strategists in the second.]

When (3.5.4) and the proof of Lemma 3.5.3 are generalized to interior equilibria of three-species games, Attempt A translates into assuming at least one of

$$x \cdot (Ax + By + Cz) \tag{3.6.2}$$
$$y \cdot (Dx + Ey + Fz) \tag{3.6.3}$$
$$z \cdot (Gx + Hy + Iz) \tag{3.6.4}$$

is negative for all nonzero $(x, y, z) \in X^1 \times Y^1 \times Z^1$ where Y^1 and Z^1 are defined as the analogue of $X^1 = \{(x_1, x_2) \mid x_1 + x_2 = 0\}$. Since X^1, Y^1, and Z^1 are all one-dimensional, each of the nine inner products in (3.6.2) to (3.6.4) is a scalar multiple of the product of the first components of the two vectors (e.g. $(x, -x) \cdot B(y, -y) = bxy$ for some $b \in R$). That is, (3.6.1) may be taken as a 3×3 matrix such as

$$\begin{bmatrix} -1 & 0 & -3 \\ -3 & -1 & 0 \\ 0 & -3 & -1 \end{bmatrix}. \tag{3.6.5}$$

In fact, given the interior equilibrium and (3.6.5), the full 6×6 matrix (3.6.1) is uniquely determined. Furthermore, for (3.6.5), at least one of (3.6.2) to (3.6.4) is negative. [To see this, suppose (3.6.2) and (3.6.3) are nonnegative. That is, assume $-x^2 - 3xz \geq 0$ and $-3xy - y^2 \geq 0$. If $x > 0$, then $z < 0$ and $y \leq 0$. In particular, $-3yz - z^2 < 0$ and so

(3.6.4) is negative. The same conclusion follows in the case $x < 0$. Finally, if $x = 0$, then $y = 0$ and thus (3.6.4) is negative unless z is also zero.]

In summary, any payoff matrix (3.6.1) with $S^* = T^* = R^* = (1/2, 1/2)$ that restricts to (3.6.5) on $X^1 \times Y^1 \times Z^1$ satisfies Attempt A. On the other hand, it is not difficult to see, by the same method used to prove Theorem 3.5.2, that the relevant linearization of the pure-strategy dynamic produces a positive multiple of (3.6.5). Since the eigenvalues, $\{-4, 1/2 \pm 3\sqrt{3}/2\, i\}$, do not all have negative real part, (S^*, T^*, R^*) is an unstable equilibrium of the pure-strategy dynamic. In particular, strong stability need not hold at equilibria satisfying Attempt A.

REMARK 3.6.1. The existence of stability conditions for competitive equilibria in economic game theory involving more than two commodities has proved elusive as well. The condition corresponding to Attempt A goes by many names in mathematical economics (e.g. perfect stability, Hicksian economy, P-matrix). [The matrix (3.6.5) is an example from Hofbauer and Sigmund (1988) given in their discussion of P-matrices.] Perhaps the most relevant connection is to the controversy surrounding the concept of perfect stability (Hicks 1939). Samuelson first conjectured (Samuelson 1941) perfect stability implied dynamic stability but later (Samuelson 1944) provided a counterexample. In fact, he showed perfect stability was neither necessary nor sufficient for dynamic stability. Since then, a great deal of research has been done to determine additional conditions on a Hicksian economy that ensure dynamic stability (e.g. the restriction that all goods be gross substitutes (Arrow and Hahn 1971)). The biological significance (Hofbauer and Sigmund 1988) of such conditions in frequency-dependent models deserves more attention.

So far, this section has considered the dynamic consequences of the obvious game-theoretic extension of Definitions 2.4.2 and 3.4.1 to multi-species systems. The next attempt, Attempt B, will be based on a dynamical-systems perspective that is again best understood at an interior equilibrium (S^*, T^*, R^*).

For a two-species interior equilibrium, Definition 3.4.1 is equivalent to (see Definition 3.2.2) the following three conditions for all nonzero $x \in X^m$ and $y \in Y^n$.

$$x \cdot Ax < 0 \qquad (3.6.6)$$
$$y \cdot Dy < 0 \qquad (3.6.7)$$
$$x \cdot Ax\, y \cdot Dy - x \cdot By\, y \cdot Cx > 0. \qquad (3.6.8)$$

Moreover, the extra condition

$$x \cdot Ax + y \cdot Dy < 0 \qquad (3.6.9)$$

can clearly be appended without affecting the aforementioned equivalence. Readers familiar with the two Routh-Hurwitz criteria (Pielou 1977), that determine whether all eigenvalues of a

2×2 matrix have negative real part, will recognize (3.6.8) and (3.6.9) as the determinant and trace criteria respectively for the matrix

$$\begin{bmatrix} x \cdot Ax & x \cdot By \\ y \cdot Cx & y \cdot Dy \end{bmatrix}. \tag{3.6.10}$$

Moreover, the negative-definiteness of A and D on X^m and Y^n respectively given in (3.6.6) and (3.6.7) are the Routh-Hurwitz criteria for the proper principal submatrices, $[x \cdot Ax]$ and $[y \cdot Dy]$, of (3.6.10).

Thus, an interior two-species ESS is characterized by asserting all principal submatrices of (3.6.10) (the matrix (3.6.10) itself is one such submatrix) satisfy the Routh-Hurwitz stability criteria for any choice of nonzero x and y in X^m and Y^n respectively. Attempt B then characterizes interior three-species ESS's by the Routh-Hurwitz criteria for all principal submatrices of

$$\begin{bmatrix} x \cdot Ax & x \cdot By & x \cdot Cz \\ y \cdot Dx & y \cdot Ey & y \cdot Fz \\ z \cdot Gx & z \cdot Hv & z \cdot Iz \end{bmatrix}. \tag{3.6.11}$$

I feel Attempt B is better than Attempt A due to the following facts, stated without proof. First, any system satisfying Attempt B automatically satisfies Attempt A. Second, a strongly stable interior equilibrium $(S^*, T^* R^*)$ must satisfy Attempt B. Third, a monomorphic population at $(S^*, T^* R^*)$ satisfying Attempt B is noninvadable by another monomorphism. Finally, if there are at most two pure strategies in each species, Attempt B implies the polymorphic pure-strategy model is l.a.s. at $(S^*, T^* R^*)$. [The above example shows this is not the case for Attempt A.]

However, Attempt B is still inadequate since there are equilibria satisfying its conditions that are not dynamically stable. By the preceding paragraph, a counterexample must involve at least three pure strategies in one of the species and at least two for the other two species. This minimum requirement is met exactly by the following counterexample that has three strategies for the first species.

As in (3.6.5), the linearized pure-strategy dynamic at an interior equilibrium $(S^*, T^* R^*)$ is given, with respect to a basis of $X^2 \times Y^1 \times Z^1$, by a 4×4 matrix L such as

$$L = \begin{bmatrix} -1-\varepsilon & 0 & 1 & 0 \\ 0 & -1-\varepsilon & 0 & 1 \\ 0 & 2 & -1 & -1/4 \\ -2 & 0 & 0 & -1 \end{bmatrix}. \tag{3.6.12}$$

Here ε is a (small) positive constant chosen so that L has eigenvalues with positive real part.

To see this is possible, note that, when $\varepsilon = 0$, the eigenvalues λ of L satisfy

$$\lambda^4 + 4\lambda^3 + 6\lambda^2 + 7/2\,\lambda + 9/2 = 0.$$

One of the Routh-Hurwitz criteria (Pielou 1977) on the coefficients of this characteristic equation is that $7/2\,((4)(6) - 7/2) > 9/2(4)^2$. Since this inequality is not true, any choice of ε sufficiently close to zero will suffice.

Thus, the pure-strategy dynamic is unstable at $(S^*,\ T^*\ R^*)$. On the other hand, if ε is any positive constant, then L satisfies the conditions of Attempt B. To see this, fix nonzero $x = (x_1,\ x_2)$, y and z in X^2, Y^1 and Z^1 respectively (with respect to the basis for L). Then

(i) L is a negative definite on X^2, Y^1 and Z^1 separately. [For instance,

   $(x,\ 0,\ 0)\cdot L(x,\ 0,\ 0) = (-1 - \varepsilon)(x_1^2 + x_2^2) < 0.]$

(ii) L satisfies the three conditions equivalent to (3.6.8). [For instance, when y is ignored,

   $(x,\ 0,\ 0)\cdot L(x,\ 0,\ 0)\,(0,\ 0,\ z)\cdot L(0,\ 0,\ z) - (x,\ 0,\ 0)\cdot L(0,\ 0,\ z)\,(0,\ 0,\ z)$
   $\cdot L(x,\ 0,\ 0)$ equals $z^2(x_1 + x_2)^2 + \varepsilon z^2(x_1^2 + x_2^2)$ which is positive.]

(iii) All eigenvalues of the matrix

$$\begin{bmatrix} -(1+\varepsilon)\left(x_1^2 + x_2^2\right) & x_1\,y & x_2 z \\ 2x_2 y & -y^2 & -yz/4 \\ -2x_1 z & 0 & -z^2 \end{bmatrix}.$$

corresponding to (3.6.11) have negative real part. This is because the characteristic equation here,

$$\lambda^3 + c_1\lambda^2 + c_2\lambda + c_3 = 0,$$

satisfies all three Routh-Hurwitz criteria for 3×3 matrices ($c_1 > 0$, $c_3 > 0$ and $c_1 c_2 > c_3$) since

$$c_1 = (1 + \varepsilon)(x_1^2 + x_2^2) + y^2 + z^2,$$

$$c_2 = y^2(x_1 - x_2)^2 + z^2(x_1 + x_2)^2 + \varepsilon\,(x_1^2 + x_2^2)(y^2 + z^2) + y^2 z^2,$$

$$c_3 = y^2 z^2\,(1/2 x_1^2 + x_2^2 + \varepsilon(x_1^2 + x_2^2)).$$

REMARK 3.6.2. The approach of this section is quite different from most of the text. One obvious difference is that the examples were chosen, as in Section 2.8, for mathematical convenience without reference to biological relevance. The practical significance of this material for biological systems is that it clearly illustrates the danger of transferring single and two-species results to multi-species systems. The other (more subtle?) difference is that, up until this section, I have made a conscious, though unannounced, effort to convince you that my approach to theoretical evolutionary game models is the "correct one". The danger in exposing the above unsuccessful attempts at ESS conditions in multi-species selection models is that it may undermine this carefully-nurtured credibility. However, I feel the ultimate importance of game-theoretic principles in these models will far outweight such personal considerations. In this context, I hope the ideas of this section and all of Chapter 3 have provided a firm base on which to develop further theoretical research.

To end this section, I should briefly mention one other possible definition for three-species ESS's. Attempts A and B were considered inadequate since neither guarantees stability of an equilibrium (for the pure-strategy dynamic). On the other hand, stability will hold when (3.4.3) is generalized. At interior equilibria, (3.4.3) is equivalent to assuming (3.6.10) and its two principal submatrices are negative-definite for any nonzero x and y in X^m and Y^n respectively. The corresponding definition for (3.6.11) implies dynamic stability; however, there are strongly stable equilibria that do not satisfy negative-definiteness. For this reason and for those mentioned in the discussion following (3.4.3), I do not view this last possibility as a fruitful approach to evolutionary stability in biology.

4. FREQUENCY-DEPENDENT EVOLUTION IN A RANDOMLY-MATING DIPLOID SPECIES

The main cause for skepticism concerning the biological relevance of evolutionary game theory as developed in the last two chapters is the implied asexual reproduction of the population. This criticism is overcome in the present chapter by assuming throughout that individuals in the large diploid population mate randomly and that each mating pair produce the same number of offspring (i.e. equal fecundity) in 1:1 sex ratio through Mendelian segregation. I feel strongly that the results contained here will convince all readers of the importance of ESS theory in population genetics.

The initial success of evolutionary game theory applied to diploid species (namely, natural selection models) is summarized in the next section. Although the results are not new to population geneticists, the pervasive game-theoretic interpretation opens the door to many exciting possibilities such as those analyzed in the rest of the chapter. In particular, ESS theory provides an intuitive geometric explanation for much of the complexity involved in frequency-dependent single-locus models (Sections 4.2 to 4.4).

4.1 Natural Selection as an Evolutionary Game

For most of this chapter, we assume that selection occurs at a single autosomal locus through different viabilities (i.e. survival to maturity) of the genotypes. In general, the viability of a genotype may depend on the frequencies of all genotypes but, for this section, natural selection will refer to the frequency-independent model.

As a basic example, assume there are exactly two possible alleles, A_1 and A_2, at the locus in question. Let $p_i(t)$ be the frequency of allele A_i in the mature population at generation t and w_{ij} be the (nonnegative) viability of genotype $A_i A_j$. If an individual's viability is independent of its sex, then the discrete-generation selection dynamic is

$$p_1(t+1) = p_1(t) \frac{w_{11} p_1(t) + w_{12} p_2(t)}{w_{11} p_1^2(t) + 2w_{12} p_1(t) p_2(t) + w_{22} p_2^2(t)}.$$ (4.1.1)

This formula can be found in most introductory texts for population genetics as well as in Hofbauer and Sigmund (1988) and Roughgarden (1979). The coincidental fact that this dynamic has already been examined in the frequency-dependent models of Chapter 2 will be exploited first.

(A) The Frequency-Dependent Haploid Perspective

Our main interest here is not the derivation of (4.1.1) but its comparison with the single-species haploid ESS theory of Chapter 2. To this end, note that (4.1.1) can be rewritten to match the discrete dynamic (2.2.6); namely,

$$p_1(t+1) = p_1(t)\frac{e_1 \cdot Wp(t)}{p(t) \cdot Wp(t)} \qquad (4.1.2)$$

where $e_1 = (1, 0)$, $p = (p_1, p_2)$ and $W = \begin{bmatrix} w_{11} & w_{12} \\ w_{21} & w_{22} \end{bmatrix}$ with $w_{12} = w_{21}$. That is,

viability selection at a diallelic locus is represented as a pure-strategy haploid evolutionary game when the population's state is given by the gene frequencies and the symmetric payoff matrix is the 2×2 selection matrix W. In this representation, the two alleles (A_1 and A_2) are the pure strategists (e_1 and e_2 respectively). In reality, the "competition" in this game occurs through the effect of random mating on these alleles.

Since there are only two pure strategies, the dynamic (4.1.2) can be analyzed completely and the results translated into the language of population genetics. For instance, when W is selectively neutral (i.e. all of the viabilities are the same: $w_{11} = w_{21}$ and $w_{12} = w_{22}$), the population remains at the initial gene frequencies and the genotypes A_1A_1, A_1A_2 and A_2A_2 are in Hardy-Weinberg proportions p_1^2, $2p_1p_2$ and p_2^2 respectively. This represents a neutrally stable equilibrium of (4.1.2). In all other cases, the population will approach a l.a.s.

If there is over-dominance (i.e. the heterozygote A_1A_2 is more viable than either of the homozygotes A_1A_1 or A_2A_2), the population approaches the polymorphism $p^* = (p_1^*, p_2^*)$ given by

$$p^* = \frac{1}{2w_{12} - w_{11} - w_{22}}(w_{12} - w_{22}, \; w_{12} - w_{11}). \qquad (4.1.3)$$

Otherwise, the population becomes monomorphic. With under-dominance, (i.e. both of the homozygotes are more viable than A_1A_2) the final monomorphism depends on the initial gene frequencies (specifically, $p_1 > p_1^*$ implies the population tends to A_1A_1). Finally, for incomplete dominance (i.e. w_{12} between w_{11} and w_{22}), evolution is to the homozygote with the largest viability. A comparison of these cases with the characterization of ESS's for the 2×2 matrix W in Section 2.4 (B) reveals the above population genetic results are consequences of two-strategy haploid ESS theory.

The straightforward generalization of (4.1.1) and (4.1.2) to the multi-allele viability-selection model given by

$$p_i(t+1) = p_i(t)\frac{e_i \cdot Wp(t)}{p(t) \cdot Wp(t)} \qquad (4.1.4)$$

suggests ESS theory applies here as well. [W is now an $n \times n$ symmetric matrix if there are n possible alleles at the single locus.] There are two reasons to proceed cautiously with this application, both discussed in Chapter 2 for general (non-symmetric) payoff matrices. The first is that the stability properties of ESS's for the discrete dynamic are not well-understood when there are more than two pure strategies (Section 2.7). The second involves the possible existence of l.a.s. equilibria of the pure-strategy dynamic that are not ESS's (Section 2.6).

In Chapter 2, the first reason for caution led to considering only the continuous dynamic. However, the transition from the discrete dynamic (4.1.4) to its continuous analogue

$$\dot{p}_i = p_i(e_i - p) \cdot Wp \qquad (4.1.5)$$

is more controversial for natural selection models than in Chapter 2. This may be because, with overlapping generations, the assumption underlying (4.1.1) that offspring genotypes are in Hardy-Weinberg proportions is not strictly valid. In this context, mathematically cumbersome concepts of "weak selection" or "short generation time" are often invoked to justify the approximation (4.1.5).

In my opinion, the controversy continues for a more technical reason; namely, that the symmetry of W can be used to characterize the stability of all equilibria of (4.1.4). Specifically, as we will see by Theorem 4.1.1, symmetry implies the mean fitness

$$\bar{w}(p) = p \cdot Wp \qquad (4.1.6)$$

is a strict Liapunov function for the discrete dynamic. [The non-trivial proof of this fundamental theorem of natural selection (Fisher 1930) for the discrete dynamic can be found in Kingman (1961) or Roughgarden (1979).] Moreover, Theorem 4.1.1 also shows that (4.1.4) and (4.1.5) have the same equilibria and each equilibrium has the same stability properties in either dynamic. Thus, there is no mathematical incentive to approximate the biologically-appealing model based on discrete generations by its continuous analogue. [A note on notation is in order here. Traditionally, the selection matrix for the continuous dynamic has entries m_{ij} (the Malthusian fitness parameter) in place of w_{ij}. We will continue using w_{ij}.]

THEOREM 4.1.1. The mean fitness function (4.1.6) is strictly increasing for both dynamics (4.1.4) and (4.1.5) unless the system is at equilibrium.

PROOF. (in the continuous case)

$$\frac{d}{dt}\bar{w}(p) = \dot{p} \cdot Wp + p \cdot W\dot{p}$$

$$= 2\sum_i \dot{p}_i e_i \cdot Wp$$

$$= 2\sum_i p_i(e_i - p) \cdot Wp \, e_i \cdot Wp \qquad \text{from (4.1.5)}$$

$$= 2\sum_i p_i(e_i - p) \cdot Wp \, (e_i - p) \cdot Wp \qquad \text{since } \sum_i p_i e_i = p$$

$$= 2\sum_i p_i[(e_i - p) \cdot Wp]^2$$

$$\geq 0$$

Furthermore $\dfrac{d}{dt}\bar{w}(p) = 0$ iff $\dot{p}_i = 0$ for all i. ∎

Classical theoretical results on natural selection also eliminate the second reason for caution. Indeed, a more careful analysis of the proof of Theorem 4.1.1 allows one to determine the direction and strength of natural selection at non-equilibrium points (see the adaptive topography approach used by Roughgarden, (1979)) but, from our ESS perspective, it is enough to note that the theorem implies l.a.s. equilibria must correspond to isolated local maxima of $\bar{w}(p)$ defined on the simplex Δ^n. Then, by Theorem 4.1.2 below, all l.a.s. equilibria are ESS's. That is, for symmetric payoff matrices, there is no need to consider mixed strategies and/or strong stability (Section 2.6) to characterize evolutionary stability.

THEOREM 4.1.2. The following three statements are equivalent.

(a) p^* is an ESS for the selection matrix W.

(b) p^* is l.a.s. for both dynamics (4.1.4) and (4.1.5).

(c) p^* is an isolated local maximum of the mean fitness function (4.1.6).

PROOF.   Let us first show the equivalence of (a) and (c). Assume W is symmetric. Then

$$p^* \cdot Wp^* - p \cdot Wp = (p^* \cdot Wp - p \cdot Wp) + (p^* \cdot Wp^* - p \cdot Wp^*). \quad (4.1.7)$$

If p^* is an ESS, $p^* \cdot Wp^* \geq p \cdot Wp^*$ by Definition 2.4.1(i). Furthermore, by Definition 2.4.2, $p^* \cdot Wp > p \cdot Wp$ for all p different from p^* in some neighborhood of p^* in Δ^n. From (4.1.7), p^* is an isolated local maximum of $\bar{w}(p)$.

Conversely, if p^* is an isolated local maximum, then the directional derivative of $p \cdot Wp$ at p^* in the direction from p^* to any other $p_0 \in \Delta^n$ is nonpositive. Since this directional derivative is $(p_0 - p^*) \cdot Wp^* + p^* \cdot W(p_0 - p^*) = 2(p_0 - p^*) \cdot Wp^*$, we have $p_0 \cdot Wp^* \leq p^* \cdot Wp^*$ for all $p_0 \in \Delta^n$. Also, if $p_0 \cdot Wp^* = p^* \cdot Wp^*$, then

$$(p^* - p_0) \cdot W(p_0 - p^*) = p^* \cdot Wp_0 - p_0 \cdot Wp_0 = p^* \cdot Wp^* - p_0 \cdot Wp_0.$$

This expression is positive if p_0 is sufficiently close (but not equal) to p^*. But the sign of $(p^* - p_0) \cdot W(p_0 - p^*)$ does not depend on this proximity and so, for such p_0, $p^* \cdot Wp_0 > p_0 \cdot Wp_0$. That is, p^* is an ESS by Definition 2.4.1.

The equivalence of statement (b) to the other two is now apparent from Theorem 4.1.1.
∎

Theorems 4.1.1 and 4.1.2 are the heart of the classical theory that views natural selection as a process whereby a population evolves to increase its mean fitness. The evolutionary outcome is then a population where all alleles present are equally fit. Moreover, the population is stable if and only if there is no nearby allele frequency-distribution with a higher mean fitness. The correspondence between ESS's of W and l.a.s. populations is an added bonus that justifies the introduction of a game-theoretic perspective into natural selection models.

(B) *The Frequency-Independent Diploid Perspective*

The temptation at this stage in the discussion of natural selection is to adopt wholeheartedly the terminology of haploid evolutionary game theory. Thus, one can talk about selfish genes (Dawkins 1976) who play strategies in and obtain payoffs through direct competition with other alleles. [Up until now, it has implicitly been individuals (in this case, genotypes) who used strategies and competed against other individuals.] The danger in this proposed approach is that it obscures the fact that the viability selection model of (4.1.4) is actually frequency independent since an individual's fitness does *not* depend on the frequency of other individuals in the population. In reality, evolution in (4.1.4) results from an interplay between random mating and the constant selection coefficients.

In hindsight, what appeared to most researchers as a lucky coincidence connecting natural selection to haploid evolutionary game theory (Akin 1979) may have been in fact unfortunate in that it discouraged them from seeking game-theoretic explanations for stability in other selection models of evolutionary biology that include such realistic effects as frequency dependence, density dependence or recombination among multiple loci. The purpose of the rest of this section is to introduce a bona fide frequency-dependent selection model and then to reinterpret, in this new context, the above results in the restricted case of natural selection. In so doing it is hoped that the mathematical complexities of this more sophisticated model can be put on an intuitive geometric level.

The general single-locus setup has all individuals of genotype $A_i A_j$ adopting the same mixed strategy (also called their phenotype) S_{ij} in Δ^m. Different genotypes may use the same strategy as, for example, $S_{12} = S_{11}$ in the case of a two-allele model with a recessive gene A_2. As before, assume 1:1 sex ratio of offspring, random mating, large population size, and sex-independent viability. [Note that these are now phenotypic viabilities.] If there are n possible alleles (there is no a priori relationship between n and m) and $p_i(t)$ is the frequency of allele A_i in the mature individuals of generation t, their offspring will initially be in Hardy-Weinberg proportions with mean strategy $S(p) = \sum_{i,j=1}^{n} p_i p_j S_{ij} \in \Delta^m$. If A is the $m \times m$ sex-independent payoff matrix whose entries a_{ij} measure the net viability gain to an individual using the ith pure strategy in a contest against the jth pure strategy (nonlinear payoffs and contests against the field can also be considered as in Cressman (1988a)), then the frequency of allele A_i in the population after selection is $p_i(t+1) = p_i(t) \sum_j p_j S_{ij} \cdot AS/S \cdot AS$. This will be rewritten as

$$p_i(t+1) = p_i(t) \frac{S^i \cdot AS}{S \cdot AS} \tag{4.1.8}$$

where $S^i = \sum_j p_j S_{ij}$ is called the *effective strategy* of allele A_i (Hines and Bishop 1984; Cressman 1988a) due to the similarity between (4.1.8) and (2.7.1). The effective strategy can be interpreted as the average strategy of those individuals in the population having allele A_i.

Suppose selection is frequency independent for the rest of this section. That is, suppose $S_{ij} \cdot AS$ depends only on S_{ij}, not on S. With $S_{ij} \cdot AS = w_{ij}$, $S \cdot AS$ and $S^i \cdot AS$ are given by $\sum_{i,j=1}^{n} p_i p_j w_{ij} = p \cdot Wp$ and $\Sigma p_j w_{ij} = e_i \cdot Wp$ respectively. Since (4.1.8) is then the same as the dynamic (4.1.4), natural selection is a frequency-independent diploid model. That is, in general, the dynamic (4.1.8) is a reasonable way to allow for actual frequency dependence in the viability selection model.

In fact, it is always possible to build abstractly a diploid model of frequency-independent natural selection that uses only two pure strategies. To see this, define $M = \max w_{ij}$ and $m = \min w_{ij}$ as the largest and the smallest of the genotypic viabilities respectively and let $A = \begin{bmatrix} m & m \\ M & M \end{bmatrix}$ be the payoff matrix. [We assume $M > m$ to avoid neutral selection.] If S_{ij} is defined to be $\left(\dfrac{\cdot}{M - m}, \dfrac{\cdot}{M - m} \right) \in \Delta^2$, then

$$S_{ij} \cdot AS = S_{ij} \cdot (m, M) = m\left(\frac{M - w_{ij}}{M - m} \right) + M\left(\frac{w_{ij} - m}{M - m} \right) = w_{ij}.$$

The "phenotype" S_{ij} of genotype $A_i A_j$ in this mathematical construction seems unconnected to biological reality. However, the heuristic information of the following theorem, gained from a comparison of natural selection at a diallelic locus with ESS theory for the matrix A, provides invaluable insight into the general dynamic (4.1.8) of frequency-dependent evolution in a diploid species.

For natural selection in the above abstract setting, the payoff matrix has ESS $S_0^* = (0, 1)$ and the mean strategy is a map from $p \in \Delta^n$ to $S(p) \in \Delta^2$. At a diallelic locus (i.e. $n = 2$), both p and $S(p)$ can be represented uniquely by their second component in the unit interval $[0, 1]$. Thus, the mean strategy map is a (quadratic) function of the unit interval into itself (see Figure 4.1).

THEOREM 4.1.3. Under natural selection at a diallelic locus, the population approaches a mean strategy $S^* = S(p^*)$ that is as close to the ESS $S_0^* = (0, 1)$ as possible given the local genetic constraints of the population. That is, p^* is a l.a.s. equilibrium of (4.1.1) if and only if, for all other $p \in \Delta^2$ in some neighborhood of p^*, $S(p)$ is further than S^* from S_0^*.

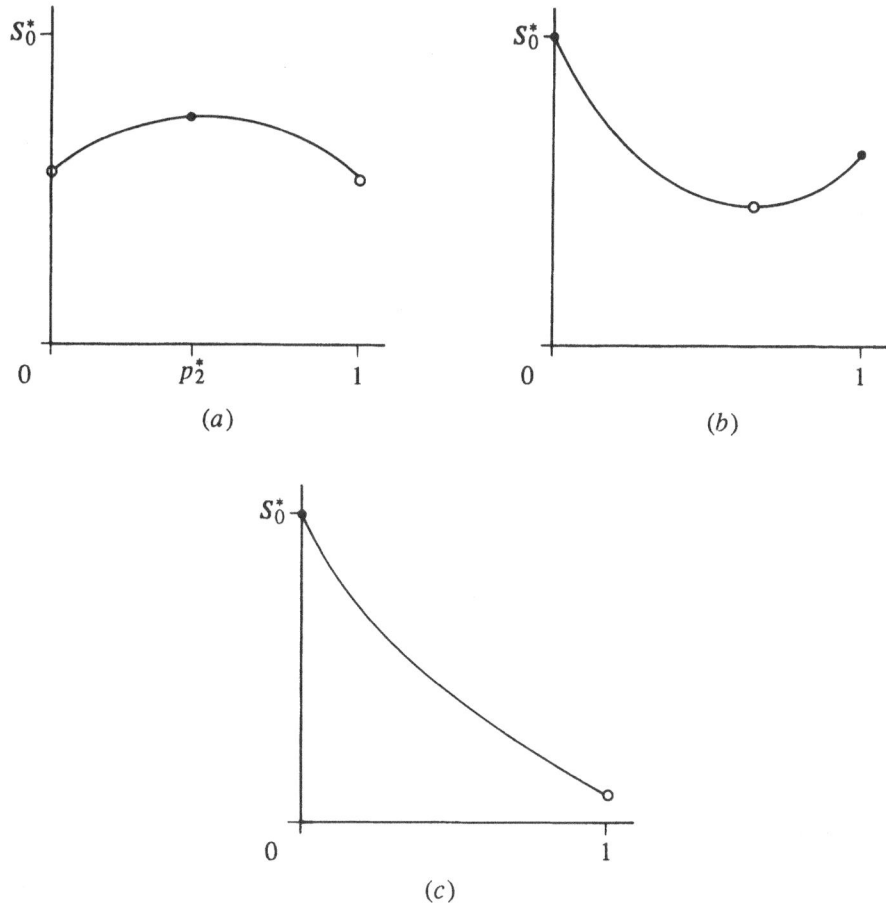

Figure 4.1. *Natural selection at a diallelic locus*

The mean strategy map (4.1.9) from Δ^2 to Δ^2 is represented by their second components with (a) over-dominance, (b) under-dominance and (c) incomplete dominance. The local maxima at the solid dots represent mean strategies that are locally as close to the height of the ESS S_0^* as possible. These are the l.a.s equilibria of natural selection. The local minima at the empty dots are unstable equilibria.

PROOF. We may assume $w_{11} \geq w_{22}$ without loss of generality. The three cases to consider are (a) over-dominance; (b) under-dominance; and (c) incomplete dominance. The mean strategy maps p_2 to the second component of $S(p)$ which is

$$(1 - p_2)^2 \frac{w_{11} - m}{M - m} + 2(1 - p_2)p_2 \left(\frac{w_{12} - m}{M - m} \right) + p_2^2 \left(\frac{w_{22} - m}{M - m} \right)$$

$$= (1/(M - m)) \left[p_2^2(w_{11} + w_{22} - 2w_{12}) + 2p_2(w_{12} - w_{11}) + (w_{11} - m) \right] \quad (4.1.9)$$

Case (a): $(M = w_{12} > w_{11} \geq w_{22} = m)$ Here the downwards parabola (4.1.9) has vertex

$$p_2^* = \frac{w_{12} - w_{11}}{2w_{12} - w_{11} - w_{22}}$$ between 0 and 1. This is the closest $S(p)$ gets to S_0^* and

corresponds to the polymorphism (4.1.3).

Case (b): $(M = w_{11} \geq w_{22} > w_{12})$ The upwards parabola again has vertex between 0 and 1. Both monomorphisms correspond to populations whose mean strategies are locally as close as possible to S_0^*. In fact, $p^* = (1, 0)$ has mean strategy S_0^*.

Case (c): $(M = w_{11} \geq w_{12} \geq w_{22})$. These viabilities are not all equal since we do not consider neutral selection. Since there is no critical point of (4.1.9) between 0 and 1, the second component of mean strategy is decreasing and the population approaches the homozygote $A_1 A_1$ corresponding to the ESS.

■

4.2 Single-Locus Models

Discrete-dynamic models of frequency-dependent viability selection at a single locus have been studied extensively in the literature. [As this text focuses only on those aspects relevant for ESS theory, neither an exhaustive discussion of these models nor a complete set of references is given here (Gayley and Michod 1990).] It has long been known that the fundamental theorem of natural selection is no longer valid. Much of the theoretical work before 1970 analyzed conditions on the viability parameters for which mean fitness increases and/or searched for other population characteristics on which to base a revised fundamental theorem. In retrospect, it can be seen that the continuation of this program (Cockerham et al. 1972; Lloyd 1977; Slatkin 1979; Lessard 1984) paralleled the development of ESS theory without explicit reference to game-theoretic principles. The connection between the two approaches was first clarified (Maynard Smith 1981; Eshel 1982; Lessard 1984) when frequency dependence is described through phenotypic strategies in Δ^2 (i.e. two-phenotype models). It is much less clear in multi-phenotype models despite many attempts for a complete analysis (Hines and Bishop 1984; Cressman 1988a; Gayley and Michod 1990).

The following elementary example of frequency-dependent viability selection at a diallelic locus will serve to illustrate the above points. Since the frequency p_1 of allele A_1 determines the genotypic frequencies uniquely through the Hardy-Weinberg relationship, the selection parameters are functions of p_1. Suppose A_1 is a recessive gene and

$$w_{11}(p_1) = 3/2 - p_1^2; \quad w_{12}(p_1) = w_{22}(p_1) = 1. \tag{4.2.1}$$

From (4.1.6), the mean fitness is then given by

$$\begin{aligned}
\bar{w}(p_1) &= p_1^2 w_{11}(p_1) + 2p_1(1 - p_1)w_{12}(p_1) + (1 - p_1)^2 w_{22}(p_1) \\
&= 1 + 1/2 p_1^2 - p_1^4.
\end{aligned}$$

An exercise in differential calculus using $d\bar{w}/dp_1 = p_1(1 - 4p_1^2)$ shows that $\bar{w}$ has a global maximum, for $0 \leq p_1 \leq 1$, at $p_1 = \frac{1}{2}$. However, by substitution of (4.2.1) into the dynamic (4.1.1), if $p_1(t) = \frac{1}{2}$ then $p_1(t + 1) = 9/17$. Thus the population's mean fitness has decreased from generation t to $t + 1$. That is, the fundamental theorem of natural selection has failed in this example. Furthermore, maximizing mean fitness does not even produce an equilibrium of the dynamic, let alone a l.a.s. one as in Section 4.1.

To develop an ESS perspective for this example, we would like to express the dynamic in the form of (4.1.8) for some choice of payoff matrix A and phenotypic strategies S_{ij}. [A careful translation of the method in Cockerham et al. (1972) shows this is always possible, by taking A as a 3×3 matrix with the three pure strategies S_{11}, S_{12}, and S_{22}, whenever the functions in (4.2.1) are quadratic.] The analysis is greatly simplified here since A can be taken as the 2×2 matrix $\begin{bmatrix} 1/2 & 3/2 \\ 1 & 1 \end{bmatrix}$ with phenotypes $S_{11} = (1, 0)$ and $S_{12} = S_{22} = (0, 1)$. To verify this, note that the mean strategy map is then

$$
\begin{aligned}
S(p) &= p_1^2 S_{11} + 2p_1(1 - p_1)S_{12} + (1 - p_1)^2 S_{22} \\
&= (p_1^2, \ 1 - p_1^2).
\end{aligned} \tag{4.2.2}
$$

Thus the components of $AS = (3/2 - p_1^2, 1)$ match (4.2.1) and so $S_{ij} \cdot AS$ has the correct form.

From Chapter 2, the only ESS of A is $S_0^* = (\frac{1}{2}, \frac{1}{2})$. The model leading to Theorem 4.1.3 suggests the population will evolve in such a way that p_1 monotonically approaches an equilibrium p_1^* that satisfies $S(p_1^*) = S_0^*$. From (4.2.2.), $p_1^* = 1/\sqrt{2}$. By looking ahead to Theorem 4.3.1, we see this is indeed the case. In summary, frequency-dependent selection in this example forces the population to adopt an ESS mean strategy rather than the strategy that maximizes the population's mean fitness.

Our purpose in the next three sections is to build, one stage at a time, the ESS perspective for models that first include multiple alleles, then multiple phenotypes and finally multiple loci. The reader is forewarned that the complete analysis is formidable compared to that of the above example and so some details are overlooked. To temper these complications somewhat, we will switch to the continuous-dynamic analogue of (4.1.8) throughout; namely,

$$
\dot{p}_i = p_i(S^i - S) \cdot AS. \tag{4.2.3}
$$

This is done for two specific reasons. First, the results in Section 4.3 for two-phenotype models remain true for (4.1.8) but are hard to prove. Also, for more than two phenotypes, the discrete stability analysis has all of the additional difficulties vis-à-vis (4.2.3) that were encountered in Section 2.7.

4.3. Two-Phenotype, Frequency-Dependent Evolution at a Single Locus

The example of Section 4.2 shows that mean fitness may not evolve monotonically in these models. However, another population characteristic, the population's mean strategy, does evolve monotonically and can be used as a basis for the following revised fundamental theorem. [Note that S is in the one-dimensional simplex Δ^2 and so its monotonic evolution refers to a particular direction on this line segment. See Lessard (1984) for the discrete proof.]

THEOREM 4.3.1. The evolution of the mean strategy is strictly monotonic for both dynamics (4.1.8) and (4.2.3) unless the system is at equilibrium.

PROOF. (in the continuous case) It is enough to show that, for any 2×2 payoff matrix A, there is a T^* on the line containing Δ^2 such that $(S - T^*) \cdot (S - T^*)$, the square of the distance from S to T^*, is strictly monotonic unless the system is at equilibrium. The evolution of S is then monotonically towards T^* or away from T^*.

The construction of T^* depends on the entries of $A = \begin{bmatrix} a & b \\ c & d \end{bmatrix}$. If $a - c \neq b - d$, there is a unique T^* on this line so that

$$AT^* = \lambda(1,\ 1) \tag{4.3.1}$$

for some real number λ. Otherwise, let T^* be one of the two pure strategies. Specifically, we may take

$$T^* = \begin{cases} \dfrac{1}{b-d+c-a}\,(b-d,\ c-a) & \text{if } a-c \neq b-d \\[2mm] (0,\ 1) & \text{otherwise} \end{cases} \tag{4.3.2}$$

In either case, from (4.2.3),

$$\frac{d}{dt}(S - T^*) \cdot (S - T^*) = 2\dot{S} \cdot (S - T^*)$$

$$= 4 \sum_{ij} \dot{p}_i p_j S_{ij} \cdot (S - T^*) \tag{4.3.3}$$

$$= 4 \sum_i p_i [(S^i - S) \cdot AS]\,[S^i \cdot (S - T^*)]$$

$$= 4 \sum_i p_i [(S^i - S) \cdot AS]\,[(S^i - S) \cdot (S - T^*)]$$

since $\sum_i p_i S^i = S$.

For monotonicity it is enough to show the terms $[(S^i - S) \cdot AS]\,[(S^i - S) \cdot (S - T^*)]$ in the last sum are either all nonnegative or else all nonpositive. To see this is independent of i, note that the two vectors $S^i - S$ and $S - T^*$ both belong to the one-dimensional subspace $X^2 = \{x = (x_1,\ x_2) \mid x_2 = -x_1\}$. Let $S^i - S = (x,\ -x)$.

If $a - c = b - d$, then, $(S^i - S) \cdot AS = ((a - c)x, (b - d)x) \cdot S = (a - c)x$. Since $S - T^* = (y, -y)$ for some $y \geq 0$, the sign of $[(S^i - S) \cdot AS][(S^i - S) \cdot (S - T^*)]$ $= (a - c)x \, 2xy = 2(a - c)x^2y$ depends only on $a - c$ and not on i.

If $a - c \neq b - d$, the terms in question have the form, by (4.3.1),

$$[(S^i - S) \cdot A(S - T^*)][(S^i - S) \cdot (S - T^*)].$$

Since the linear transformation A reduces to scalar multiplication on X^2, the sign of each term depends only on this scalar and not on i.

Finally, for strict monotonicity, suppose all terms in the last sum of (4.3.3) are zero. Then, for each i, one of the four conditions $p_i = 0$; $(S^i - S) \cdot AS = 0$; $S^i = S$; or $S = T^*$ must hold. It is clear from (4.2.3) that any one of the first three imply $\dot{p}_i = 0$. Also, if $S = T^*$, then either $AS = \lambda(1, 1)$ by (4.3.1) or S is the extreme point $(0, 1)$ of Δ^2. In the latter case $S^i = S$ for all i and so the last condition also implies $\dot{p}_i = 0$. Thus evolution of S is strictly monotonic unless at equilibrium.

∎

Theorem 4.1.3 on natural selection at a diallelic locus is a corollary of the above result (take $A = \begin{bmatrix} m & m \\ M & M \end{bmatrix}$ and T^* the ESS $S_0^* = (0, 1)$ of A). The geometric intuition initiated there generalizes to frequency-dependent two-phenotype models. Indeed, T^* defined in (4.3.2) is closely connected to the ESS structure of A. If T^* is outside Δ^2, there is only one ESS (a pure strategy) and evolution is towards it. If $T^* \in \Delta^2$, then either it is the only ESS and S evolves towards it or else S evolves monotonically to one of the two pure-strategy ESS's.

In fact, if S^* is an attainable ESS for A given the genotypic strategies (i.e. $S(p) = S^*$ for some $p \in \Delta^n$), then the following immediate consequence of Theorem 4.3.1 is a perfect illustration of how ESS theory can predict the outcome of evolution without recourse to the particular dynamic.

THEOREM 4.3.2. Suppose S^* is an ESS for A and $L = \{p \in \Delta^n \mid S(p) = S^*\}$ is nonempty. Then S^* is l.a.s. for the mean strategy evolution induced by (4.2.3). That is, if p is initially sufficiently close to a $p^* \in L$, then $S(p)$ will evolve to S^*.

The converse of Theorem 4.3.2 is not true. For instance, natural selection at a diallelic locus depicted in Figure 4.1 (a) and (b) has non-ESS l.a.s. equilibria when there is over (or under) dominance. However, this stability (often called *internal*) depends critically on the particular genotypic strategies in use. That is, any non-ESS equilibrium can be invaded by mutant alleles if these allow the population mean strategy to move closer to the ESS. The reader should be reminded of Section 2.6 where non-ESS l.a.s. equilibria for the pure-strategy haploid dynamic can be made unstable by introducing individuals with appropriately chosen mixed strategies. The following definition generalizes the strong stability concept of Section 2.6 to diploid species. [I find the conventional phrase "evolutionary stability" used here difficult to distinguish from evolutionarily stable in Chapter 2. However, I will continue to use it since more accurate terminology such as diploid strong stability or diploid evolutionary stability is overly cumbersome.]

DEFINITION 4.3.3. S^* is said to exhibit *evolutionary stability* if, whenever S^* is an attainable population mean strategy of a diploid population, it is a l.a.s. equilibrium for the mean strategy evolution determined by (4.2.3).

By Theorems 4.3.1. and 4.3.2, evolutionary stability is a complete dynamic characterization of ESS's in this diploid model just as Theorem 2.6.2 relates strong stability to ESS's for haploid species. That is, we have

THEOREM 4.3.4. S^* has evolutionary stability for a single-locus, two-phenotype model if and only if S^* is an ESS.

An appealing picture of evolution in stages now emerges. At first, the population approaches an internally stable mean strategy S that is as close as possible to the ESS S^* given current genetic constraints. As genetic mutations occur only those that shift S towards S^* are maintained and a new internally stable state is reached. This process continues in evolutionary time until the population reaches S^*. At this point mutations can still invade but not alter the population's mean strategy. Of course, this picture assumes all mutant genotypes have strategies in the original simplex Δ^2.

The description of internal equilibria of (4.2.3) has also received much attention in the literature (see Gayley and Michod (1990) for a thorough review). The emphasis here will be on characterizing their stability by reference to previous results in Chapter 4.

From (4.2.3), it is clear that an equilibrium p^* must satisfy $S^i(p^*) = S(p^*)$ whenever $p_i^* > 0$ unless $AS^* = \lambda(1, 1)$ as in (4.3.1). The former are called genotypic equilibria (Lessard 1984) and often depend on the population's genetic structure. Any monomorphism ($p_i^* = 1$ for some i) as well as the polymorphic interior equilibria in Figure 4.1 correspond to genotypic equilibria.

DEFINITION 4.3.5. (a) A *phenotypic equilibrium* $S^* \in \Delta^2$ satisfies $AS^* = \lambda(1, 1)$ for some real number λ.

(b) A *genotypic equilibrium* $p^* \in \Delta^n$ is non-phenotypic and satisfies $S^i(p^*) = S(p^*)$ for all $p_i^* > 0$.

A phenotypic equilibrium is l.a.s. if and only if it is an ESS. The internal stability of genotypic equilibria p^* can be described by means of the local frequency-independent $n \times n$ viability matrix W^* defined componentwise by

$$w_{ij}^* = S_{ij} \cdot AS(p^*). \tag{4.3.4}$$

The following description is strikingly similar to Theorem 4.1.2 on frequency-independent natural selection.

THEOREM 4.3.6. Let $p \cdot W^* p$ be the local mean fitness function. A genotypic equilibrium p^* is internally stable (this includes neutral stability) for (4.2.3) if and only if p^* is a local maximum of this function. Moreover, p^* is l.a.s. if and only if p^* is an isolated local maximum.

PROOF. By Theorem 4.3.1 and its proof, the evolution of $S(p)$ is monotonic and p^* is stable iff, for all p sufficiently close to p^*,

$$(S(p) - S(p^*)) \cdot AS(p^*) \leq 0. \tag{4.3.5}$$

But $S(p) \cdot AS(p^*) = p \cdot W^* p$. Thus $p \cdot W^* p - p^* \cdot W^* p^* \leq 0$ is equivalent to (4.3.5) and the proof is complete.

∎

Biologically, the theorem says the population is stable when it maximizes mean fitness according to the existing frequency-independent selection coefficients (a local version of the fundamental theorem of natural selection). This intuitive result is a direct consequence of the assumption that individuals do not make rational decisions. Individual selection produces a myopic population that can only change according to local evolutionary pressures. There is no group selection force to shift the mean strategy fitness to higher levels.

[The existence of polymorphic genotypic equilibria is related biologically to heterosis. In particular Cressman (1988a) showed, if all heterozygotes are incompletely dominant (i.e. w_{ij} is between w_{ii} and w_{jj} for all i and j), then no such equilibria exist. The practical significance of Theorem 4.3.6 is limited by the necessity of determining p^* before calculating W^* as opposed to the reverse process in Theorem 4.1.2. We encountered the same difficulty when applying matrix ESS theory in Section 2.9 to nonlinear fitness functions.]

4.4. Multi-Phenotype, Frequency-Dependent Evolution at a Single Locus

At the time this text was written, there is no known population characteristic on which to base a fundamental theorem of frequency-dependent viability selection when the population has more than two phenotypes. This section is not meant to describe ongoing theoretical research in the area. Rather its purpose is to illustrate both the difficulties and insights that arise by considering the evolution of the population's mean strategy in these models. At first this is accomplished by a dynamical analysis of two viability-selection examples at a diallelic locus with three phenotypes. Stability properties of equilibria for general models are then discussed by linearization techniques. The reader may be disappointed but should not be surprised by our return to the linearized dynamic since the existence of an elementary Liapunov method for stability analysis usually yields a fundamental theorem.

(A)   Examples

For both of the examples, suppose that the genotypic strategies for the three genotypes A_1A_1, A_1A_2 and A_2A_2 are the pure strategies; $(1, 0, 0)$, $(0, 1, 0)$ and $(0, 0, 1)$ respectively. The mean strategy map from Δ^2 to Δ^3 and the effective strategy of allele A_1 are $S(p) = (p^2, 2p(1 - p), (1 - p)^2)$ and $S^1(p) = (p, 1 - p, 0)$ respectively where p is the frequency of allele A_1. Thus the set of attainable mean strategies in Δ^3 is the parabola sketched in Figure 4.4.1.

The examples differ only in their respective 3×3 payoff matrices (given in Table 4.4) that were chosen, among other reasons, to have an equilibrium when $p = 1/2$ for the one-dimensional continuous dynamic

$$\dot{p} = p(S^1(p) - S(p)) \cdot AS(p) \tag{4.4.1}$$

given by (4.2.3). Geometrically, any such polymorphic equilibrium is an attainable mean strategy S where the vector AS is perpendicular to the parabola in Figure 4.4.1.

In example (a), the payoff matrix has the two ESS's, $(0, 0, 1)$ and $(1/7, 6/7, 0)$. Here, there are four dynamic equilibria of (4.4.1); at $p = 1$, $1/2$, $1/4$ and 0. Figure 4.4.2 depicts

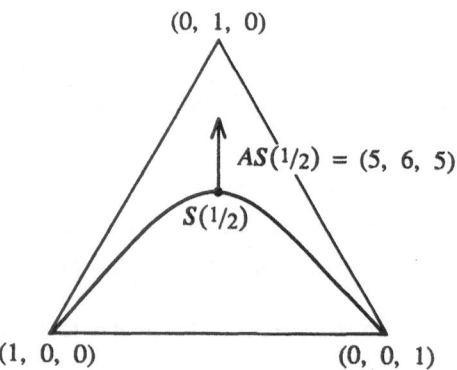

Figure 4.4.1. *The strategy simplex* Δ^3

For the two examples of Section 4.4, there is an equilibrium at $p = 1/2$ since $AS(1/2)$ is perpendicular to the parabola of attainable mean strategies.

71

Table 4.4. *Two examples of* 3 × 3 *payoff matrices*

For both examples, the dynamic (4.4.1) has an equilibrium when $p = 1/2$.
Example (b) is taken from Cressman (1988b).

$$\begin{bmatrix} 5 & 5 & 5 \\ 11 & 4 & 5 \\ 11 & 1 & 7 \end{bmatrix}$$

(a)

$$\begin{bmatrix} 5 & 5 & 5 \\ 5 & 4 & 11 \\ 5 & 1 & 13 \end{bmatrix}$$

(b)

this information on the strategy simplex as well as that for example (b) which has only one ESS, (0, 0, 1), and three equilibria at $p = 1$, $1/2$ and 0.

For the two-phenotype models of Section 4.3, the evolution of S is globally towards an ESS. This continues to be the case locally near the attainable ESS (0, 0, 1) in Figure 4.4.2 and also for the stable polymorphic equilibrium in example (a) even though $p = 1/2$ is not the closest Euclidean point on the parabola to $(1/7, 6/7, 0)$. However, example (b) shows that distance to S^* cannot serve globally as the base for a fundamental theorem since S evolves away from S^* when $p > 1/2$.

We will discuss the local evolution near an ESS later. For now, consider the polymorphic equilibrium $p = 1/2$ in the two examples. An alternative geometric explanation of its stability

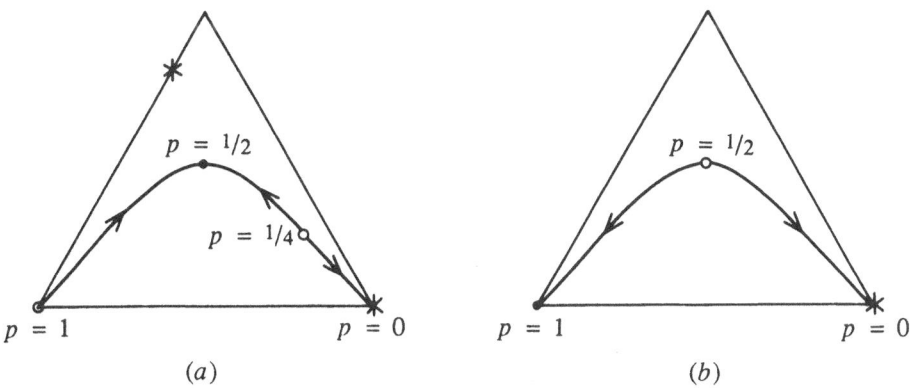

Figure 4.4.2. *The ESS structure for Table 4.4 matrices*

ESS's are denoted by asterisks (*). The dynamic flow of (4.4.1) is indicated by arrows for both payoff matrices; stable equilibria by solid dots; and unstable equilibria by empty dots.

was initiated by Hines and Bishop (1984) for multi-phenotype models when there is an interior ESS. Their method suggests that, since selection favors those strategies in the direction of $AS(p)$, $p = 1/2$ should be stable in the situation depicted in Figure 4.4.1. In other words, since $AS(1/2)$ is on the opposite side of the tangent line to the set of attainable mean strategies, the population's genetic constraints will not allow further evolution towards more favorable strategies. The method fails in example (b), perhaps because the ESS is on the boundary (see Theorem 4.4.1 below for an alternative explanation).

In the rest of this section, we will develop a generalization of Hines and Bishop's method to include payoff matrices with no interior ESS as in Table 4.4. To avoid some technicalities, only stability at polymorphic equilibria will be considered. Finally, the l.a.s. of attainable interior ESS's will be discussed.

(B)   Linearization

Suppose $p^* \in \Delta^n$ is a polymorphic equilibrium of (4.2.3) and $S^* = S(p^*)$ is in the interior of Δ^m. Let $p_i = p_i^* + x_i$ where $\Sigma x_i = 0$. From (4.2.3),

$$(S^{*i} - S^*) \cdot AS^* = 0. \qquad (4.4.2)$$

where $S^{*i} = \sum_j p_j^* S_{ij}$ for $i = 1, 2, \cdots , n$ and

$$\dot{x}_i = \dot{p}_i = (p_i^* + x_i) \left(\sum_j (p_j^* + x_j) S_{ij} - \sum_{jk} (p_j^* + x_j)(p_k^* + x_k) S_{jk} \right)$$
$$\cdot A \sum_{jk} (p_j^* + x_j)(p_k^* + x_k) S_{jk}$$

$$= (p_i^* + x_i)(S^{*i} - S^*) \cdot AS^*$$

$$+ p_i^* \left[\left(\sum_j x_j S_{ij} - 2 \sum_j x_j S^{*j} \right) \cdot AS^* + 2(S^{*i} - S^*) \cdot A \sum_j x_j S^{*j} \right] + o(|x|).$$

Use (4.4.2) and $\Sigma x_j = 0$ to simplify this linearization to

$$\dot{x}_i = p_i^* \left[\Sigma x_j S_{ij} \cdot AS^* + 2(S^{*i} - S^*) \cdot A \Sigma x_j S^{*j} \right] + o(|x|). \qquad (4.4.3)$$

To analyze the stability of p^*, one can consider the $n \times n$ matrix associated to (4.4.3) as was done in Section 2.8 (Cressman 1988a). However, for our purposes, it is sufficient to consider the function

$$V(p) = 1/2 \sum_{i=1}^{n} x_i^2 / p_i^* . \qquad (4.4.4)$$

From (4.4.3),

$$\dot{V}(p) = \Sigma x_i \dot{x}_i / p_i^* = \Sigma x_i x_j S_{ij} \cdot AS^* + 2 \Sigma x_i S^{*i} \cdot A \Sigma x_j S^{*j} + o(|x|^2). \qquad (4.4.5)$$

Since $S - S^* = 2 \sum_j x_j S^{*j} + \sum_{ij} x_i x_j S_{ij}$ and $\Sigma x_j S^{j*} \cdot AS^* = 0$, the quadratic terms of $\dot{V}(p)$ will be nonpositive if the following two inequalities hold.

$$(S - S^*) \cdot AS^* \leq 0 \qquad\qquad\qquad (4.4.6\,\text{a})$$

$$\Sigma x_i S^{*i} \cdot A \Sigma x_j S^{*j} \leq 0. \qquad\qquad\qquad (4.4.6\,\text{b})$$

Both inequalities are biologically and mathematically significant. From (4.3.5), the first is equivalent to p^* maximizing the local mean fitness function $p \cdot W^* p$. It is also equivalent to the Hines and Bishop (1984) requirement that AS^* be on the opposite side of the tangent plane to the local set of attainable mean strategies. The second inequality asserts that A is negative semi-definite on the subspace of X^m spanned by the vectors $\{S^{*i} - S^* \mid i = 1, \cdots, n\}$ that is parallel to this same tangent plane. That is, A satisfies the ESS conditions when restricted to this subspace (see Theorem 2.4.5 c).

In fact, if $\{S^{*1}, \cdots, S^{*n}\}$ is linearly independent, the argument following (2.8.6) shows that $V(p)$ is a Liapunov function for the dynamic (4.2.3) when both inequalities in (4.4.6) are true. In particular, p^* is l.a.s. in this case.

For both of the examples earlier in this section, $\{S^{*1}, S^{*2}\}$ is linearly independent and (4.4.6 a) is true at $p^* = 1/2$. The crucial difference at this equilibrium is that A is negative definite in the first example where p^* is stable but not in the second where p^* is unstable.

In general, the vectors S^{*i} may be linearly dependent. The dynamic (4.2.3) then contains a centre manifold at p^* that determines the stability of S^*. The analysis on this centre manifold is substantially more difficult than in Section 2.8. Necessary and sufficient stability criteria are not known. However, if the centre manifold has the same dimension as $\{p \in \Delta^n \mid S(p) = S^*\}$, then it consists entirely of equilibria as in Section 2.8 (B) and we have the following generalization of Hines and Bishop's method.

THEOREM 4.4.1. If p^* is a local maximum of W^* and A is negative definite on $\{\Sigma x_i S^{*i} \mid \Sigma x_i = 0\}$, then S^* is l.a.s. for the mean strategy dynamic determined by (4.2.3).

For instance, if A has an interior ESS (not necessarily at S^*), then (4.4.6 b) is automatically satisfied and one needs only check (4.4.6 a). On the other hand, if S^* is an interior ESS, both inequalities are automatic. This provides strong evidence that an attainable interior ESS is l.a.s. for any multi-phenotype single-locus model (i.e. that an ESS exhibits evolutionary stability). Over the years I have alternated between believing this result and not. It now appears to be false due to a three-phenotype, three-allele counterexample referred to by Hines (1991). [By Theorem 4.3.4, at least three phenotypes are needed and it is fairly easy to show at least three alleles are needed as well.] The biological relevance of such counterexamples remains open to debate but there can be no question that the appealing theoretical picture of evolution stopping when mutations have pushed the population to an ESS (Hammerstein 1990) is not completely accurate.

The converse of Theorem 4.4.1 is not true. As an example, let A be the negative of the matrix in Table 4.4 (b). Then $p^* = 1/2$ is l.a.s at the same time as p^* is a global minimum of W^*. Stability follows here since $V(p)$ is still a Liapunov function ($\dot{V}(p) \leq 0$) even though one of its quadratic terms is positive.

Finally, the other polymorphic equilibrium, $p^* = 1/4$, in example (b) deserves mention. Since $AS^* = AS(1/4) = 5(1, 1, 1)$, this equilibrium is also called a phenotypic equilibrium after Definition 4.3.5. It is unstable solely because the inequality in (4.4.6 b) is opposite to what it should be.

4.5. A Two-Locus, Two-Allele, Two-Phenotype Example

Just as mean fitness does not necessarily increase for single-locus, frequency-dependent selection models, it has long been known that the fundamental theorem of frequency-independent natural selection is not valid in general two-locus viability models (Bodmer and Felsenstein 1967). Since natural selection can always be placed in a two-phenotype setting (Section 4.1 (B)), it follows that the mean strategy need not evolve monotonically in two-locus models (i.e. Theorem 4.3.1. is not valid here). This section first develops the general model in game-theoretic terms and then analyzes a specific example where Theorem 4.3.1 fails.

Suppose each locus has two alleles; A_1 and A_2 at the first, B_1 and B_2 at the other. Number the four gamete types A_1B_1, A_1B_2, A_2B_1 and A_2B_2 as 1, 2, 3 and 4 respectively with frequencies p_1, p_2, p_3 and p_4 before selection. An ij-individual's strategy $S_{ij} \in \Delta^2$ is assumed to depend only on the genes present at the two loci and not on how they are paired on the chromosomes or on the individual's sex. Then the double heterozyzotes, $A_1B_1A_2B_2$ and $A_1B_2A_2B_1$, are indistinguishable with the same strategy $S_{14} = S_{23}$.

With random mating after selection and 1:1 sex ratio of offspring, the discrete dynamic becomes

$$p_i(t + 1) = \frac{p_i S^i \cdot AS + c_i rD S_{14} \cdot AS}{S \cdot AS} \tag{4.5.1}$$

where $S^i = \sum_{j=1}^{4} p_j S_{ij}$ (the effective strategy of gamete i) and $S = \Sigma p_i S^i$ (the mean strategy) are evaluated at generation t and A is the 2×2 payoff matrix. The only change from the single-locus dynamic (4.1.8) is the additional term $c_i rD S_{14} \cdot AS$ that reflects the fact the transmission of gamete types by double heterozygotes may depend on recombination. From Roughgarden (1979, chapter 8) or Michod (1984), we have $c_2 = c_3 = -c_1 = -c_4 = 1$; D is the linkage disequilibrium $p_1p_4 - p_2p_3$; and r is the constant recombination fraction (between 0 and $1/2$).

If there is tight linkage (e.g. the two loci are adjacent on the same chromosome), then there is no recombination (i.e. $r = 0$) and (4.5.1) reduces to a two-phenotype single-locus model with four gametes. Thus, without recombination, the fundamental theorem stating evolution is monotonic towards an ESS is valid. The following example assumes the other extreme; that is, the loci are independent (e.g. on different chromosomes) and so $r = 1/2$.

Let $A = \begin{bmatrix} M & M+1 \\ M+1 & M \end{bmatrix}$ where M is some positive constant, to be specified later, that reflects the strength of selection towards the ESS, $S_0^* = (1/2, \ 1/2)$, of A. [Selection is weaker as M increases.] If $S_{ij} = (0, \ 1)$ for all genotypes except $S_{13} = (1/2, \ 1/2)$, the population mean strategy at $S^* = (1/4, \ 3/4)$ is as close as genetically possible to S_0^* only when the second locus is homozygous at B_1 (i.e. $p_2^* = p_4^* = 0$) and the heterozygote frequency at the first locus is maximized at $p_1^* = p_3^* = 1/2$.

The example describes the biological situation of a modifier at the second locus. With fixation of B_1, the population will monotonically approach the stable over-dominant polymorphism that has equal numbers of both alleles at the first locus. Heuristically, the introduction of the modifier B_2 should not alter this monotonicity since any individual with this allele is the least fit in the population (i.e. strategy $(0, \ 1)$ has the lowest individual payoff for any attainable mean strategy). We will now see this is not the case.

The mean strategy $S = (p_1 p_3, \ 1 - p_1 p_3)$ evolves towards S_0^* in one generation if and only if $p_i(t+1)$ from (4.5.1) satisfies

$$p_1(t+1)p_3(t+1) > p_1 p_3.$$

If $p_1 = 1/4$, $p_2 = 1/2$ and $p_3 = 1/4$ initially, substitution of $S^1 = (1/2 p_3, \ 1 - 1/2 p_3)$, $S^3 = (1/2 p_1, \ 1 - 1/2 p_1)$, $AS = (M + 1 - p_1 p_3, \ M + p_1 p_3)$, $S_{14} \cdot AS = M + p_1 p_3$ and $D = -p_2 p_3$ into (4.5.1) yields

$$p_1(t+1) = \frac{\frac{1}{4}\left(\frac{1}{8}, \frac{7}{8}\right) \cdot \left(M + \frac{15}{16}, \ M + \frac{1}{16}\right) - \frac{1}{2}\left(-\frac{1}{8}\right)\left(M + \frac{1}{16}\right)}{\left(\frac{1}{16}, \frac{15}{16}\right) \cdot \left(M + \frac{15}{16}, \ M + \frac{1}{16}\right)}$$

$$= \frac{\frac{1}{4}\left(M + \frac{22}{128}\right) + \frac{1}{16}\left(M + \frac{1}{16}\right)}{M + \frac{15}{128}}$$

and

$$p_3(t+1) = \frac{\frac{1}{4}\left(M + \frac{22}{128}\right) - \frac{1}{16}\left(M + \frac{1}{16}\right)}{M + \frac{15}{128}}.$$

Thus $p_1(t+1)p_3(t+1) = \dfrac{\frac{1}{16}\left(M + \frac{22}{128}\right)^2 - \left(\frac{1}{16}\right)^2\left(M + \frac{1}{16}\right)^2}{\left(M + \frac{15}{128}\right)^2}$. As M becomes

large, $p_1(t+1)p_3(t+1) \cong \dfrac{\frac{1}{16}M^2 - \left(\frac{1}{16}\right)^2 M^2}{M^2} < \dfrac{1}{16} = p_1 p_3.$

That is, for selection pressure weak enough, the initial high level of linkage disequilibrium forces the mean strategy S to begin its evolution in the opposite direction from S^*. However, ESS theory remains useful in this example in that S eventually evolves to S^* (as long as $p_1 + p_3 > 0$ initially) even though it may start in the wrong direction. Furthermore, if S is initially sufficiently close to S^*, the evolution will be monotonic since $p_1 p_3$ then becomes a Liapunov function.

As mentioned previously in this chapter, an exhaustive mathematical discussion of genetic constraints imposed on ESS theory has not been attempted. The reader need only skim the relevant chapters of Hofbauer and Sigmund (1988) or of Akin (1990) to appreciate the impossibility of such a task. However, I remain firmly convinced that evolutionary game theory is the best avenue to gain a qualitative understanding of this complex topic.

5. FREQUENCY- AND DENSITY-DEPENDENT EVOLUTION IN A HAPLOID SPECIES

Chapters 2 and 3 systematically developed the stability conditions for the frequency-dependent haploid dynamic in two biological contexts (the single-species and two-species systems respectively). Based on these heuristic conditions, Chapter 4 gave a game-theoretic interpretation to classical results in the genetic evolution of a diploid species. The present chapter assumes a new game-theoretic aspect in that individual fitness may also depend on population size (i.e density). For this reason, new stability conditions for a *density-dependent* evolutionarily stable strategy (DDESS) must be developed before their dynamic consequences are ascertained.

5.1 Frequency- and Density-Dependent Fitness (and the Haploid Dynamic)

First recall the following relevant notation for a single haploid species introduced in Chapter 2. Let $S_1, S_2, \cdots, S_n \in \Delta^m$ list all possible individual phenotypes of the species and N_i be the number of individuals of phenotype S_i at time t. Then $S = \Sigma p_i S_i \in \Delta^m$ is the time-dependent population mean strategy where $p_i = N_i/N$ and $N = \Sigma N_i$ are the frequencies of strategy S_i and the population size respectively.

In this chapter, we assume an individual's fitness not only depends on its strategy and on S but also depends on N. Specifically, let $F(S, N) = (F_1(S, N), \cdots, F_m(S, N))$ be the m-dimensional fitness vector whose real-valued k^{th} component $F_k(S, N)$ represents the fitness of the kth pure strategy when the population is in the state (S, N). We then assume an individual using (mixed) strategy $S_i \in \Delta^m$ has expected fitness $S_i \cdot F(S, N)$ and so the continuous haploid dynamic becomes

$$\dot{N}_i = N_i S_i \cdot F(S, N). \tag{5.1.1}$$

A non-trivial equilibrium of (5.1.1) is an n-dimensional point $(N_1^*, \cdots, N_n^*)$ with non-negative components that satisfies $N^* > 0$ and $S_i \cdot F(S, N) = 0$ whenever $N_i^* > 0$. In this setting, evolutionary game theory is concerned with the stability of the population state (S^*, N^*). For instance, (S^*, N^*) is l.a.s. if, whenever the population is initially close to such an equilibrium, its state evolves to (S^*, N^*). [In contrast, the frequency evolution of Chapter 2 is unconcerned with the stability of N^*.]

To simplify the stability analysis, we will assume throughout that the fitness vector has the special form

$$F(S, N) = A^* S + (N - N^*)A'S \tag{5.1.2}$$

where A^* and A' are $m \times m$ matrices. From a biological perspective, the population is engaged in a game whose *density-dependent payoff matrix* is given by

$$A(N) = A^* + (N - N^*)A'.$$

Mathematically, the matrices A^* and A' for an equilibrium state (S^*, N^*) are determined (Cressman 1988c) from the Taylor Series expansion similar to (2.9.5) for the generally nonlinear $F(S, N)$.

As before, the main purpose of the text in this new setting is to develop static conditions involving fitness comparisons that predict the long-term outcome of the evolutionary dynamic (5.1.1). To this end, the ESS program outlined in Section 2.6 will be followed by first considering monomorphic populations.

5.2 Monomorphic DDESS's and Stability

In Chapters 2 and 3, heuristic conditions for a monomorphic ESS were proposed before their usefulness for dynamic stability was proven. The process will be reversed here. That is, monomorphic DDESS conditions are formulated to ensure stability is guaranteed upon invasion by a mutant strategy. Only then are the conditions given a heuristic interpretation that is of equal importance but difficult to motivate without the preliminary groundwork.

Suppose that the population is initially near the equilibrium state (S^*, N^*) with N_1 individuals of phenotype S^* and N_2 individuals of (a different) phenotype S. In other words, the monomorphic S^* population with N_1 close to N^* is invaded by a few mutant S-individuals. From the last section, the two-dimensional continuous dynamic is

$$\dot{N}_1 = N_1 S^* \cdot (A^* \mu + (N - N^*)A'\mu)$$
$$\dot{N}_2 = N_2 S \cdot (A^* \mu + (N - N^*)A'\mu)$$

(5.2.1)

where $\mu = (N_1 S^* + N_2 S)/N$ and $N = N_1 + N_2$.

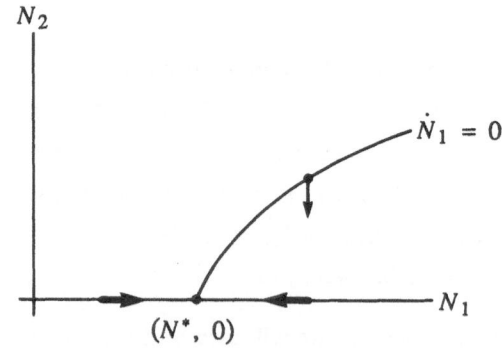

Figure 5.2. *The phase portrait of the dynamic (5.2.1)*

The N_1-isocline is a smooth curve that is not horizontal at the equilibrium $(N^*, 0)$. For l.a.s. at $(N^*, 0)$, the flows must be in the directions indicated by the arrows.

(S^*, N^*) resists this invasion if and only if $(N_1, N_2) = (N^*, 0)$ is l.a.s. for (5.2.1). In particular, when $N_2 = 0$, N^* must be a l.a.s. equilibrium for the one-dimensional population dynamic

$$\dot{N}_1 = N_1 S^* \cdot (A^* S^* + (N_1 - N^*)A'S^*).$$

The flow along the N_1-axis of Figure 5.2 has the correct direction if and only if

$$S^* \cdot A'S^* < S^* \cdot A^* S^* = 0. \tag{5.2.2}$$

Since $S^* \cdot A'S^* < 0$, the N_1-isocline given by

$$S^* \cdot A^*\mu + (N - N^*)S^* \cdot A'\mu = 0 \tag{5.2.3}$$

is not tangent to the N_1-axis at $(N^*, 0)$. Stability requires $\dot{N}_2 < 0$ along this isocline (Figure 5.2). Solve (5.2.3) for $N - N^*$ and substitute this into $\dot{N}_2$ from (5.2.1) to obtain

$$S \cdot A^*\mu - \frac{S^* \cdot A^*\mu \, S \cdot A'\mu}{S^* \cdot A'\mu} < 0 \tag{5.2.4}$$

for (N_1, N_2) near $(N^*, 0)$. Since $S^* \cdot A'\mu$ is negative for $\mu = \varepsilon \, S + (1 - \varepsilon)S^*$ near S^*,

$$S \cdot A^*\mu \, S^* \cdot A'\mu - S^* \cdot A^*\mu \, S \cdot A'\mu > 0. \tag{5.2.5}$$

Expansion of (5.2.5) in powers of ε implies $S \cdot A^*S^* \leq 0$. Furthermore, if $S \cdot A^*S^* = 0$, then

$$S \cdot A^*S \, S^* \cdot A'S^* - S^* \cdot A^*S S \cdot A'S^* \geq 0 \tag{5.2.6}$$

and if (5.2.6) is an equality, $S \cdot A^*S \, S^* \cdot A'S - S^* \cdot A^*S S \cdot A'S > 0$.

To avoid some unnecessary mathematical complications, we will assume that (5.2.6) is never an equality (when $S \neq S^*$). [This regularity assumption is not explicitly stated in what follows.] From the above discussion, a monomorphic DDESS must then satisfy

DEFINITION 5.2.1. (S^*, N^*) is a *monomorphic DDESS* if

(i) $S^* \cdot A^*S^* = 0$ $\qquad\qquad$ (5.2.7)

(ii) $S^* \cdot A'S^* < 0$ $\qquad\qquad$ (5.2.8)

and, for all individual strategies S different from S^*,

(iii) $S \cdot A^*S^* \leq 0$ $\qquad\qquad$ (5.2.9)

(iv) $S \cdot A^*S \, S^* \cdot A'S^* - S^* \cdot A^*S \, S \cdot A'S^* > 0$ if (iii) is an equality. $\qquad$ (5.2.10)

Comparison with Definition 2.2.1 shows (5.2.7) and (5.2.9) combine to yield (2.2.2) while (5.2.10) is similar to (2.2.3). The remaining density condition (5.2.8) has no counterpart in Chapter 2 where only the frequency dynamic is considered. The following version of Theorem 2.2.2 justifies the above definition.

THEOREM 5.2.2. (S^*, N^*) is a monomorphic DDESS if and only if it is l.a.s. for the dynamic (5.2.1) and any choice of individual strategy S different from S^*.

PROOF. Necessity has already been shown. To prove sufficiency, assume that (S^*, N^*) is a monomorphic DDESS. Linearization of (5.2.1) using $N_1 = N^* + x_1$ and $N_2 = x_2 \geq 0$ produces

$$\dot{x}_1 = S^* \cdot A^* S x_2 + N^* S^* \cdot A'S^* (x_1 + x_2) + o(|x_1| + |x_2|)$$

$$\dot{x}_2 = x_2 \left[S \cdot A^* S^* + S \cdot A^*(S - S^*) \frac{x_2}{N^*} \right. \tag{5.2.11}$$

$$\left. + S \cdot A'S^* (x_1 + x_2) + o(|x_1| + |x_2|) \right].$$

The eigenvalues of the linearized system are $N^* S^* \cdot A'S^*$ and $S \cdot A^* S^*$. If $S \cdot A^* S^* < 0$, both eigenvalues are negative and l.a.s. follows.

If $S \cdot A^* S^* = 0$, there is a one-dimensional centre manifold. By the change of variables, $y = S^* \cdot A^* S x_2 + N^* S^* \cdot A'S^* (x_1 + x_2)$ and $x = x_2$, the dynamic (5.2.11) has the same form as (2.10.1); namely,

$$\dot{x} = x \left(x \frac{S \cdot A^* S}{N^*} + S \cdot A'S^* \left(\frac{y - S^* \cdot A^* Sx}{N^* S^* \cdot A'S^*} \right) + o(|x| + |y|) \right)$$

$$\dot{y} = N^* S^* \cdot A'S^* y + o(|x| + |y|).$$

By Section 2.10, this centre manifold has the form $y = o(|x|)$ and so the relevant dynamic (2.10.2) is

$$\dot{u} = u^2 \left[\frac{S \cdot A^* S}{N^*} - \frac{S \cdot A'S^* S^* \cdot A^* S}{N^* S^* \cdot A'S^*} \right] + o(|u|^2).$$

Since $u \geq 0$, (S^*, N^*) is l.a.s. if $S \cdot A^* S S^* \cdot A'S^* - S \cdot A'S^* S^* \cdot A^* S > 0$ (i.e. if (S^*, N^*) satisfies the last monomorphic DDESS condition). [It is interesting to note here that if equality is allowed in (5.2.6), a higher order expansion is needed in (5.2.11) to verify stability.]
∎

Definition 5.2.1 is the most convenient method to verify (S^*, N^*) is a monomorphic DDESS when fitness has the form (5.1.2). However, its heuristic value is diminished by the apparent lack of biological justification for condition (5.2.10). This can be remedied by developing an equivalent definition through the introduction of the surface along which the average per capita growth rate is zero. From (5.1.1) or (5.2.1), this density isocline where $\dot{N} = 0$ must satisfy $\mu \cdot F(\mu, N) = 0$. From (5.1.2), for μ sufficiently close to S^*, there is a unique density $N = \phi(\mu)$ given by

$$\phi(\mu) = N^* - \frac{\mu \cdot A^* \mu}{\mu \cdot A' \mu}$$

for which per capita growth is zero. To summarize we have

DEFINITION 5.2.3. The smooth positive-valued function ϕ defined on Δ^m is given implicitly by

$$\phi(\mu) = N \quad \text{if and only if} \quad \mu \cdot F(\mu, N) = 0.$$

In particular, $\phi(S^*) = N^*$ and ϕ is well-defined for all μ sufficiently close to S^*.

Definition 5.2.1 has the following heuristic form as is proven in Theorem 5.2.5.

DEFINITION 5.2.4. (S^*, N^*) is a *monomorphic DDESS* if, for all individual strategies different from S^*, the following two conditions hold whenever $\mu = \varepsilon S + (1 - \varepsilon)S^*$ for some sufficiently small positive ε.

(i) $S^* \cdot F(\mu, \phi(\mu)) > \mu \cdot F(\mu, \phi(\mu)) = 0$ (5.2.12)

(ii) $\dfrac{\partial}{\partial N} \mu \cdot F(\mu, N) < 0$ if N is sufficiently close to N^*.   (5.2.13)

Both conditions have biological significance. The density condition, (5.2.13), asserts that at fixed density the per capita growth rate decreases as population size increases. If, in addition, S^* does better along the density isocline (condition (5.2.12)) than the mutant strategy, then S cannot invade successfully.

As an aside, many biologists would rather have condition (5.2.12) replaced by the condition $S^* \cdot F(\mu, N^*) > \mu \cdot F(\mu, N^*)$ that only involves frequency effects. If this were so, then one could verify stability by first fixing density (apply Definition 2.4.2) and then fixing frequency (condition (5.2.13)). This approach, considered by Cressman and Dash (1987), is unwarranted as evident from the examples in the next section. The same problem arises in ecological contexts (Riechert and Hammerstein 1983) and has, to a large extent, been ignored. Although not pursued in this text, I feel evolutionary game theory has much more to offer in such ecological models.

THEOREM 5.2.5. Definitions 5.2.1 and 5.2.4 are equivalent.

PROOF. From (5.1.2), $\dfrac{\partial}{\partial N} \mu \cdot F(\mu, N) = \dfrac{\partial}{\partial N} (\mu \cdot A^*\mu + (N - N^*)\mu \cdot A'\mu) = \mu \cdot A'\mu.$ Thus (5.2.13) is equivalent to $\mu \cdot A'\mu < 0$ for ε sufficiently small positive. By taking $\mu = S^*$ in Definition 5.2.4 we obtain $S^* \cdot A^* S^* = 0$ and $S^* \cdot A'S^* < 0$ from the two conditions.

For general μ, (5.2.12) becomes

$$S^* \cdot (A^*\mu + (\phi(\mu) - N^*)A'\mu) = S^* \cdot A^*\mu - \frac{\mu \cdot A^*\mu \, S^* \cdot A'\mu}{\mu \cdot A'\mu} > 0.$$

Multiplication by the negative expression $\mu \cdot A'\mu$ and substitution of $\varepsilon S + (1 - \varepsilon)S^*$ for μ both times it appears on the left side of an inner product produces

$$\varepsilon [S^* \cdot A^*\mu \, S \cdot A'\mu - S \cdot A^*\mu \, S^* \cdot A'\mu] < 0.$$

The last inequality is a reformulation of (5.2.5). The discussion following (5.2.5) completes the proof.

◼

5.3 The Density-Dependent Hawk-Dove Game

In Chapter 2, the per capita growth rate of the population did not depend on the density of the haploid species. In particular, for the hawk-dove example of Section 2.3 with payoff matrix

$$A = \begin{bmatrix} -1 & 2 \\ 0 & 1 \end{bmatrix}$$ (i.e. with resource value $V = 2$ and fighting cost $C = 4$), frequencies for

the pure-strategy dynamic evolve to the ESS $S_0^* = (1/2, \ 1/2)$ while population size approaches an exponential growth rate of $S_0^* \cdot A S_0^* = 1/2$.

A more realistic interpretation of the above model is to regard A as representing only that part of individual fitness due to direct competition between individuals. Another term is then included in an individual's fitness, that is independent of its strategy, to reflect the biological intuition that population growth rates decrease as density increases. Rowe et al. (1985) and Cressman et al. (1986) refer to this factor as the *background fitness* of the species. For instance, the density-dependent payoff matrix

$$A(N) = \begin{bmatrix} -1 - N & 2 - N \\ -N & 1 - N \end{bmatrix}$$ (5.3.1)

models the above hawk-dove game at $N = 0$ and has the same density effect on each payoff entry. [For mathematical convenience, I have chosen a matrix that exhibits drastic density effects for small changes in N. For added realism, N may be considered as measuring population size in thousands (or millions!).]

As expected, the description of monomorphic DDESS's for (5.3.1) will depend critically on the list of possible individual strategies. However, as seen in the next section, for 2×2 matrices of the form (5.3.1), an individual strategy S^* is the first component of a monomorphic DDESS (S^*, N^*) under invasion by both pure strategists if and only if (S^*, N^*) is a DDESS. The straightforward, though instructive, determination of these DDESS's follows. If S^* is a pure strategy, the corresponding diagonal entry of (5.3.1) must be zero by (5.2.7). Thus $S^* = (0, 1)$ and $N^* = 1$ defines a possible DDESS. That it is not follows from (5.2.9) since $S \cdot A(N^*)S^* = S \cdot (1, 0) > 0$ for any $S \neq (0, 1)$.

Thus S^* must be a mixed strategy for any DDESS. By (5.2.9), $A^*S^* = (0, 0)$. In particular, the determinant $(-1 - N^*)(1 - N^*) + N^*(2 - N^*) = 2N^* - 1$ of $A(N^*)$ is zero. That is, $N^* = 1/2$, $A^* = \begin{bmatrix} -3/2 & 3/2 \\ -1/2 & 1/2 \end{bmatrix}$ and $S^* = (1/2, \ 1/2)$. The only condition of Definition 5.2.1 that is not obviously true is (5.2.10). Since $S \cdot A'S^* = S^* \cdot A'S^* = -1$, for all $S \in \Delta^m$, (5.2.10) is equivalent to

$$S \cdot A^*S < S^* \cdot A^*S. \tag{5.3.2}$$

This inequality holds since S^* is an interior ESS of A^* (see (2.4.2)). In summary, $(S_0^*, N_0^*) = ((1/2, 1/2), 1/2)$ is the only DDESS of (5.3.1).

From this example, it is apparent a simpler method can be used to find DDESS's when density has the same negative effect on each payoff. First determine the ESS's, S_0^*, of $A(0)$. For each such S_0^*, determine N_0^* such that $S_0^* \cdot A(N_0^*)S_0^* = 0$. If N_0^* is positive, then (S_0^*, N_0^*) is a DDESS. Conversely any DDESS is of this form.

Monomorphic DDESS's can be found by combining the above simple method with the discussion in Section 2.3. For (5.3.1), any mixed-strategy model that includes individuals using strategy $S_0^* = (1/2, 1/2)$ has $((1/2, 1/2), 1/2)$ as its only monomorphic DDESS. Moreover, any other mixed-strategy model has a monomorphic DDESS if and only if all individual strategies are on the line segment joining S_0^* to one of the pure strategies and the individual strategy S^* nearest S_0^*, say $S^* = (p^*, 1 - p^*) \in \Delta^2$, has corresponding positive N^* where $N^* = 1 - 2(p^*)^2$. That is, if $p^* < 1/\sqrt{2}$, the only monomorphic DDESS is at (S^*, N^*) and, if $p^* \geq 1/\sqrt{2}$ there is none. In this latter case, the population goes extinct.

Evolutionary game theory as presented in this text is also fundamentally concerned with stability for the polymorphic dynamic (5.1.1). With payoff matrix (5.3.1), the dynamic has the same form as (2.8.1) by taking $A = A(0)$ and $\rho(N) = -N$. Then, by Lemma 2.8.1, the DDESS, $S_0^* = (1/2, 1/2)$ and $N_0^* = 1/2$, is l.a.s. if S_0^* is a convex combination of the initial individual strategies $S_1, \cdots, S_n$. In fact, it is not hard to show this interior DDESS is globally stable in this case.

Thus, for this example, there is no essential difference between ESS theory from Chapter 2 and DDESS theory. The model can be viewed as another means to increase the situations in which ESS theory applies. Alternatively, I view the above discussion as a means to validate the frequency approach of ESS theory *when density has the same effect on each payoff entry*.

The rest of this section examines the correspondence between ESS theory and DDESS theory when density effects differ from strategy to strategy. The two subsections analyze, in turn, two density-dependent payoff matrices for their DDESS structure and their stability in the pure-strategy dynamic.

(A) Non-Aggressive Behavior Enhanced at High Density

Perhaps it is most realistic to assume density has a greater effect on Hawk fitness than on Dove. That is, given the same population mean strategy, assume the aggressive Hawk behavior is more adversely affected by a crowded environment that may inhibit the effectiveness of aggression. For illustrative purposes, take

$$A(N) = \begin{bmatrix} -1 - 3N & 2 - 3N \\ -2N & 1 - 2N \end{bmatrix}. \qquad (5.3.3)$$

This models the same hawk-dove game as (5.3.1) at low density but includes a factor that decreases Hawk fitness fifty percent faster than Dove as density increases.

If either pure strategy can invade, the method used earlier in this section shows there are only the following two possibilities for monomorphic DDESS's.

(i) $N^* = 1/2$ and $S^* = (0, 1)$ (5.3.4)

(ii) $N^* = 1/3$ and $S^* = (1/3, 2/3)$. (5.3.5)

By (5.2.9), the first possibility is not a DDESS. For an interior DDESS in a two-strategy game, inequality (5.2.10) is again the pivotal condition and this need only be verified when S is a pure strategy, say e_1. Notice that, for (5.3.5),

$$A^* = \begin{bmatrix} -2 & 1 \\ -2/3 & 1/3 \end{bmatrix} \quad \text{and} \quad A' = \begin{bmatrix} -3 & -3 \\ -2 & -2 \end{bmatrix}. \qquad (5.3.6)$$

Thus, $(S_0^*, N_0^*) = ((1/3, 2/3), 1/3)$ is the only (monomorphic) DDESS for (5.3.3) because (5.2.10) is the positive expression

$$e_1 \cdot A^* e_1 \, S_0^* \cdot A' S_0^* - S_0^* \cdot A^* e_1 \, e_1 \cdot A' S_0^* = (-2)(-7/3) - (-10/9)(-3) = 4/3.$$

The analysis of dynamic stability for (S_0^*, N_0^*) under (5.1.1) is not so straightforward as it was for (5.3.1). For example, the pure-strategy dynamic with $S_1 = e_1$ and $S_2 = e_2$ (and $N = N_1 + N_2$) becomes

$$\begin{aligned}
\dot{N}_1 &= N_1((-1 - 3N)N_1 + (2 - 3N)N_2)/N \\
\dot{N}_2 &= N_2((-2N)N_1 + (1 - 2N)N_2)/N.
\end{aligned} \qquad (5.3.7)$$

The two non-trivial equilibria (N_1^*, N_2^*) are the points $(0, 1/2)$ and $(1/9, 2/9)$ given by (5.3.4) and (5.3.5) respectively. The 2×2 Jacobian matrix $\begin{bmatrix} \partial \dot{N}_1/\partial N_1 & \partial \dot{N}_1/\partial N_2 \\ \partial \dot{N}_2/\partial N_1 & \partial \dot{N}_2/\partial N_2 \end{bmatrix}$ of the linearized dynamic (evaluated at each equilibrium) determines local stability in both cases. By elementary calculus the Jacobians are

$$\begin{bmatrix} 1/2 & 0 \\ -2 & -1 \end{bmatrix} \quad \text{and} \quad \begin{bmatrix} -1 & 0 \\ -8/9 & -2/9 \end{bmatrix}$$

with eigenvalues $\{1/2, -1\}$ and $\{-1, -2/9\}$ respectively. Thus, (5.3.5) corresponds to a l.a.s. equilibrium of (5.3.7) while (5.3.4) does not.

The conclusion could have again been predicted by ESS theory since $S_0^* = (1/3, 2/3)$ is an ESS of A^* given in (5.3.6) whereas $S^* = (0, 1)$ is not an ESS of $A(1/2)$. At this stage it seems (5.2.10) can be replaced by the simpler condition (5.3.2) (see the aside discussed after Definition 5.2.4). The simplification does not work in general as we will now see.

(B) Aggressive Behavior Enhanced at High Density

Readers familiar with human reactions to crowded environments that often bring out aggressive behavior (e.g. rush hour in urban areas) may consider the payoff matrix

$$\begin{bmatrix} -4 - N & 6 - N \\ -7 - 3N & 13 - 3N \end{bmatrix} \tag{5.3.8}$$

reflects more realistic density effects than (5.3.3). [Although Dove does better against itself than against Hawk and Hawk does worse against itself, (5.3.8) does not represent a typical hawk-dove game at low density. However, we will continue to use the terminology that interprets the first row of (5.3.8) as payoffs to Hawks.]

A short exercise shows the only possible DDESS has

$$N_0^* = 1 \quad \text{and} \quad S_0^* = (1/2, 1/2) \tag{5.3.9}$$

with corresponding $A^* = \begin{bmatrix} -5 & 5 \\ -10 & 10 \end{bmatrix}$ and $A' = \begin{bmatrix} -1 & -1 \\ -3 & -3 \end{bmatrix}$. In fact, this point is a monomorphic DDESS since

$$e_1 \cdot A^* e_1 \, S_0^* \cdot A' S_0^* - S_0^* \cdot A^* e_1 \, e_1 \cdot A' S_0^* = (-5)(-2) - (-15/2)(-1) = 5/2$$

is positive. This is somewhat surprising since S_0^* is *not* an ESS of A^*.

An examination of dynamic stability for (5.3.9) causes even more confusion. On the one hand, (S_0^*, N_0^*) is l.a.s. under invasion by any single mutant strategy S (see Theorem 5.2.2). On the other hand, the pure-strategy dynamic

$$\dot{N}_1 = N_1 \left((-4 - N)N_1 + (6 - N)N_2 \right)/N$$
$$\dot{N}_2 = N_2 \left((-7 - 3N)N_1 + (13 - 3N)N_2 \right)/N \tag{5.3.10}$$

is unstable at the corresponding equilibrium $(N_1^*, N_2^*) = (1/2, 1/2)$ since the two eigenvalues of the Jacobian matrix $1/2 \begin{bmatrix} -6 & 4 \\ -13 & 7 \end{bmatrix}$ sum to $1/2$ and so cannot both have negative real part.

We are faced with the following predicament. If a polymorphic DDESS is to be defined as an (S^*, N^*) that is a monomorphic DDESS for all other choices of $S \in \Delta^m$ (STEP 3 in Section 2.6) then DDESS's are not necessarily stable (for the pure-strategy dynamic). It is thus tempting to add the further DDESS condition that S^* be an ESS for A^* and so discard possibilities such as (5.3.9). The proposal works beautifully for the examples considered so far and, at the same time, suggests the strong stability concept may also be extended to density-dependent games. That we must proceed cautiously with the general theory in the next section is apparent from the following example.

Consider the density-dependent payoff matrix

$$A(N) = \begin{bmatrix} -3 - 2N & 7 - 2N \\ -4 - 6N & 16 - 6N \end{bmatrix}. \qquad (5.3.11)$$

Comparison with (5.3.8) shows that this model has the same monomorphic DDESS; namely (5.3.9). [In fact, (5.3.8) and (5.3.11) are members of the one-parameter family of matrices $\begin{bmatrix} -5 - (N-1)k & 5 - (N-1)k \\ -10 - 3(N-1)k & 10 - 3(N-1)k \end{bmatrix}$ all of which have the same monomorphic DDESS but different density effects.] The only difference near this equilibrium is that (5.3.11) has double the density effect of (5.3.8). Whatever static (polymorphic) DDESS conditions are chosen, it seems clear that if (5.3.9) is a DDESS for one model it must be for the other. We have already seen that (5.3.9) is not strongly stable for (5.3.8). Unfortunately, this causes a problem for any DDESS definition since (5.3.9) *is* strongly stable for (5.3.11). [For instance, the pure-strategy dynamic has Jacobian matrix $1/2 \begin{bmatrix} -7 & 3 \\ -16 & 4 \end{bmatrix}$ whose eigenvalues both have real part $-3/4$.] In particular, a monomorphic DDESS can imply stability of the pure-strategy dynamic even if S_0^* is not an ESS of A^*.

Before attempting to sort out these difficulties in the next section, I should comment on the fact that all my examples have had the same density effect on a given individual irrespective of the frequency of other strategies (i.e. all rows of A' are negative multiples of the vector $(1, 1)$). There is no need for this. In fact, there is no a priori reason all entries of A' are negative. Indeed, for an altruistic strategy such as warning conspecifics of imminent danger, it seems reasonable that a crowded environment may increase this strategy's payoff by decreasing the cost of issuing such warnings.

5.4 The DDESS Conditions and Strong Stability

In my view, the highlights of Chapters 2 and 3 were the equivalence of static ESS conditions and dynamic strong stability as proven in Theorems 2.6.2 and 3.5.2 respectively. The hawk-dove examples show that some compromise must be made when density effects are taken into account. This text chooses not to alter the strong stability concept.

DEFINITION 5.4.1. (S^*, N^*) is *strongly stable* if, whenever S^* is a convex combination of $\{S_1, \cdots, S_n\}$, (S^*, N^*) is a l.a.s. equilibrium for the dynamic (5.1.1).

Any strongly stable (S^*, N^*) must be a monomorphic DDESS for any other $S \in \Delta^m$ (Theorem 5.2.2) but the converse is not true (e.g. payoff matrix (5.3.8)). For this reason, the DDESS definition will append the extra condition (5.3.2) to Definition 5.2.1. [The reader will notice the following definition allows equality in (5.3.2). Reasons for this and consequences of it are discussed after stability is shown in Theorem 5.4.3.]

DEFINITION 5.4.2. (S^*, N^*) is a *DDESS* if the following equivalent statements, (a) and (b), are satisfied.

(a) (i) $S^* \cdot A^*S^* = 0$

   (ii) $S^* \cdot A'S^* < 0$

   (iii) If $S \in \Delta^m$ is different than S^*, then $S \cdot A^*S^* \leq S^* \cdot A^*S^*$

   (iv) If there is equality in (iii), then $S \cdot A^*S \leq S^* \cdot A^*S$ and
   $$S \cdot A^*S \ S^* \cdot A'S^* - S^* \cdot A^*S \ S \cdot A'S^* > 0.$$

(b) (i) $\dfrac{\partial}{\partial N} S \cdot F(S, N) < 0$ if (S, N) is sufficiently close to (S^*, N^*)

   (ii) $S \cdot F(S^*, N^*) \leq S^* \cdot F(S^*, N^*)$ for all $S \in \Delta^m$ different from S^*

   (iii) If there is equality in (ii), then for S sufficiently close to S^*,
   $$S \cdot F(S, N^*) \leq S^* \cdot F(S, N^*) \text{ and } S^* \cdot F(S, \phi(S)) > S \cdot F(S, \phi(S)) = 0.$$

Strong stability is not equivalent to being a DDESS (e.g. the strongly stable equilibrium (5.3.9) of payoff matrix (5.3.11) is not a DDESS). However, from the examples considered so far, it is still possible that any DDESS is strongly stable. This conjecture would then place strong stability between the DDESS and monomorphic DDESS concepts (i.e. DDESS implies strong stability implies monomorphic DDESS). The following weaker version of the conjecture (Cressman 1990) falls just short of a complete proof since the dynamic on the centre manifold is not fully analyzed (the same stumbling block occurred in Section 3.5).

THEOREM 5.4.3. Suppose $N_1^*, \cdots, N_n^*$ is an equilibrium of (5.1.1) that corresponds to a DDESS (S^*, N^*). If this equilibrium is regular (i.e. $S_i \cdot A^*S^* < S^* \cdot A^*S^*$ whenever $N_i^* = 0$), then (S, N) will evolve to (S^*, N^*) if (S, N) is initially sufficiently close to (S^*, N^*).

PROOF. To linearize the dynamic (5.1.1), $\dot{N}_i = N_i S_i \cdot (A^* + (N - N^*)A')S$, define y_i as $N_i = N_i^* + y_i$. Since $S = \Sigma N_j S_j / \Sigma N_j = (N^*S^* + \Sigma y_j S_j)/(N^* + \Sigma y_j)$,

$$\dot{y}_i = (N_i^* + y_i)S_i \cdot (A^* + \Sigma y_j A')(N^*S^* + \Sigma y_j S_j)\frac{1}{N^*}\left(1 - \frac{\Sigma y_j}{N^*} + \cdots\right)$$

$$= \Sigma_j \left(\frac{N_i^*}{N^*} S_i \cdot A^* S_j y_j + N_i^* S_i \cdot A'S^* y_j\right) + y_i S_i \cdot A^*S^* + o(|y|). \tag{5.4.1}$$

Since the equilibrium is regular, the eigenvalues $S_i \cdot A^*S^*$ corresponding to $N_i^* = 0$ are all negative. Thus, for stability analysis, we may assume that $N_i^* > 0$ for all $i = 1, 2, ..., n$. This implies that $S_i \cdot A^*S^* = 0$ and, more importantly, that

$$\Sigma x_i S_i \cdot A^* \Sigma x_j S_j \leq 0 \qquad (5.4.2)$$

for all $x \in X^n$.

The rest of the proof assumes (5.4.2) is a strict inequality for all nonzero $\Sigma x_i S_i \in X^m$ (the degenerate case allowing equality is justified in Cressman (1990) and summarized after this proof). The proof parallels that of Theorem 3.5.2 by taking an orthonormal basis of R^n, with respect to the inner product

$$\langle y, z \rangle = \Sigma \frac{y_i z_i}{N_i^*}, \qquad (5.4.3)$$

such that the last basis vector is $v^* = \dfrac{1}{N^*}(N_1^*, \cdots, N_n^*)$. The other basis vectors are then all in X^n and any $y \in R^n$ can be written uniquely in the form $x + bv^*$ for some $x \in X^n$ and $b \in R$.

Let L be the $n \times n$ matrix whose entries are defined through reference to (5.4.1) by

$L_{ij} = N_i^* \left(\dfrac{1}{N^*} S_i \cdot A^* S_j + S_i \cdot A' S^* \right)$. From (5.4.3), $\langle x + bv^*, L(\xi + \beta v^*) \rangle$ is

$$\Sigma_i \frac{1}{N_i^*} (x_i + bN_i^*/N^*) \left[(N_i^*/N^*) S_i \cdot A^* \left(\Sigma_j \xi_j S_j + \beta S^* \right) + N_i^* \beta S_i \cdot A' S^* \right]$$

$$= \frac{1}{N^*} \Sigma x_i S_i \cdot A^* \Sigma \xi_j S_j + \frac{b}{N^*} S^* \cdot A^* \Sigma \xi_j S_j + \beta \Sigma x_i S_i \cdot A' S^* + b\beta S^* \cdot A' S^*.$$

In particular

$$\langle x, L(x + bv^*) \rangle = \frac{1}{N^*} \Sigma x_i S_i \cdot A^* \Sigma x_j S_j + b \Sigma x_i S_i \cdot A' S^* \qquad (5.4.4)$$

and

$$\langle bv^*, L(x + bv^*) \rangle = \frac{b}{N^*} S^* \cdot A^* \Sigma x_j S_j + b^2 S^* \cdot A' S^*. \qquad (5.4.5)$$

For every $x \in X^n$, either (5.4.4) or (5.4.5) is negative unless $b = 0$ and $\Sigma x_i S_i = 0$. [If this were not so, then $b\Sigma x_i S_i \cdot A' S^* \geq -\dfrac{1}{N^*} \Sigma x_i S_i \cdot A^* \Sigma x_j S_j \geq 0$ by (5.4.2). From (5.2.8), $bS^* \cdot A^* \Sigma x_j S_j \geq -N^* b^2 S^* \cdot A' S^* \geq 0$. This implies

$$b^2 \Sigma x_i S_i \cdot A' S^* \ S^* \cdot A^* \Sigma x_j S_j \geq b^2 \Sigma x_i S_i \cdot A^* \Sigma x_j S_j \ S^* \cdot A' S^*$$

which can be shown to contradict (5.2.10) after a short calculation.]

A combination of Lemmas 3.5.3 and 3.5.4 together with the non-degeneracy assumption on (5.4.2) shows that all eigenvalues of L have negative real part if $\{ S_1, \cdots, S_n \}$ are linearly independent. This result and an application of centre manifold theory as used at the end of Theorem 3.5.2 completes the proof.
∎

As previously mentioned, the DDESS definition includes the possibility $x \cdot A^*x = 0$ (i.e. equality in (5.3.2)) for some nonzero $x \in X^m$. In fact, Cressman (1990) showed that, when all $N_i^* > 0$, a more precise condition for stability in Theorem 5.4.3 is to replace (5.4.2) by the requirement

$$x \cdot A^*x + N^*S^* \cdot A'S^* < 0 \qquad (5.4.6)$$

for all $x \in X^m$ such that $S^* + x \in \Delta^m$. [This weaker inequality is implied by (5.4.2) and (5.2.8).]

Mathematically, a DDESS definition that combines condition (5.4.6) with the conditions of a monomorphic DDESS is awkward since (5.4.6) is difficult to verify for a given density-dependent payoff matrix. However, this inequality does explain the different stabilities for the two hawk-dove games (5.3.8) and (5.3.11). The smaller density effect for model (5.3.8) implies there is an appropriate x_0 for which (5.4.6) is not true and this in turn implies instability of the pure-strategy dynamic. The non-existence of such a constrained x_0 implies the strong stability of model (5.3.11).

Biologically, we see that strong stability does not imply the DDESS conditions for our Definition 5.4.2 since individual strategies are restricted to lie in Δ^m (i.e. if arbitrary $x \in X^m$ were permitted, then (5.4.6) would imply $x \cdot A^*x \leq 0$ for all $x \in X^m$). This is connected to the idea of super-strategies mentioned in Section 2.4 (A). For example, if Super-Hawks are considered in model (5.3.11) with strategy $S_0^* + x_0 = (1/2, 1/2) + (1, -1) = (3/2, -1/2)$, then from (5.3.9)

$$x_0 \cdot A^*x_0 + N_0^*S_0^* \cdot A'S_0^* = 10 - 4 > 0.$$

That is, model (5.3.11) becomes unstable at (S_0^*, N_0^*) if invasion by these Super-Hawks is allowed. A more careful analysis when the population is near the stable equilibrium (S_0^*, N_0^*) shows that payoffs for individuals whose strategies are most extreme (i.e. most Hawkish or most Doveish) are higher than for individuals adopting moderate mixed strategies. This suggests mutant super-strategists will eventually appear to destabilize the system.

5.5 Density-Dependent Natural Selection as a Haploid Evolutionary Game

The first four sections of the present chapter have analyzed theoretical density-dependent evolutionary games under the assumption that the species is haploid. In fact, the dynamic (5.1.1) first appeared as a model of density-dependent natural selection for a diploid species. This section briefly develops this connection — emphasizing the game-theoretic interpretation of classical results.

Density-dependent viability selection at a single autosomal locus of a diploid population has been studied by a number of researchers. The presentation below will follow most closely the approach of Ginzburg (1977), Roughgarden (1979) and Cressman (1988c). At the same time, the presentation parallels Section 4.1 on frequency- and sex-independent selection.

Suppose $A_1, A_2, \cdots, A_n$ are the n possible alleles at the single locus and N_i is the number of gametes A_i at time t. Then $N = \Sigma N_i$ is twice the number of diploid individuals in the population. [The factor 2 is a mathematical nuisance that can be ignored in what follows.]

If the viability of genotype A_iA_j at density N is given by $w_{ij}(N)$, the continuous dynamic under random mating becomes

$$\dot{N_i} = N_ie_i \cdot W(N)p \qquad (5.5.1)$$

where $p = (p_1, \cdots, p_n)$ gives the frequencies $p_i = N_i/N$ of allele A_i and $W(N)$ is the symmetric, density-dependent, $n \times n$ selection matrix with entries $w_{ij}(N)$. We will assume throughout this section that $W(N)$ is given by

$$W(N) = W^* + (N - N^*)W' \qquad (5.5.2)$$

so that the above dynamic has the form (5.1.1) where the allelic "strategies" are the n pure strategies associated to the "payoff" matrix $W(N)$. [A short calculus exercise verifies that the frequency dynamic, $\dot{p_i} = p_i(e_i - p) \cdot W(N)p$ matches (4.1.5).]

It is the symmetry of W that again facilitates our analysis of the model. First, the static DDESS conditions are much simplified.

THEOREM 5.5.1. Suppose $p^* \in \Delta^n$ and N^* is positive. The following three statements are equivalent.
(a) (p^*, N^*) is a DDESS of $W(N)$.
(b) (p^*, N^*) is a monomorphic DDESS for arbitrary $p \in \Delta^n$.
(c) p^* is an ESS of W^* and N^* is the stable density of the monomorphic population using strategy p^*.

PROOF. If $p \cdot W^*p^* = p^* \cdot W^*p^* = 0$, then $p^* \cdot W^*p = 0$ by symmetry. Therefore, (5.2.10) is equivalent to

$$p \cdot W^*p < p^* \cdot W^*p \qquad (5.5.3)$$

(since $p^* \cdot W'p^* < 0$ by (5.2.8)). That is, Definitions 5.2.1 and 5.4.2 are equivalent (i.e. (a) $\Leftrightarrow$ (b)).

Inequalities (5.2.9) and (5.5.3) are those of Definition 2.4.1. Moreover, N^* is a stable equilibrium of

$$\dot{N} = Np^* \cdot W(N)p^* \text{ iff } p^* \cdot W^*p^* = 0 \text{ and } p^* \cdot W'p^* < 0$$

(i.e. conditions (5.2.7) and (5.2.8)). Thus (b) and (c) are equivalent.

∎

Second, there is no need to introduce mixed strategies or a strong stability concept when characterizing dynamic stability. Indeed, the following two theorems are versions of Theorem 4.1.2 for the density-dependent model.

THEOREM 5.5.2. (p^*, N^*) is a DDESS of $W(N)$ if and only if it is an isolated local maximum of the function $\phi: \Delta^n \to R^+$ (defined implicitly in Definition 5.2.3 by $p \cdot W(\phi(p))p = 0$) and $p^* \cdot W'p^* < 0$.

PROOF. By (5.5.2), $\phi(p^*) = N^*$ iff $p^* \cdot W^*p^* = 0$. Furthermore, from the definition of ϕ, $p \cdot W^*p + (\phi(p) - N^*)p \cdot W'p = 0$. Since $p^* \cdot W'p^* < 0$ holds for both statements to be proven equivalent, (p^*, N^*) is an isolated local maximum iff

$$p \cdot W^*p < p^* \cdot W^*p^* = 0 \qquad (5.5.4)$$

for all $p \in \Delta^n$ sufficiently close (but not equal) to p^*.

For any p in this neighborhood, let $p = p^* + x$, where $x \in X^n$. Then, by symmetry, $p \cdot W^*p = 2x \cdot W^*p^* + x \cdot W^*x$. By considering the linear and quadratic terms in x separately, (5.5.4) is equivalent to

(i) $\quad p \cdot W^*p^* = x \cdot W^*p^* \le 0 = p^* \cdot W^*p^*$ and

(ii) $\quad p \cdot W^*p = x \cdot W^*x < 0 = p^* \cdot W^*p$ if there is equality in (i).

Since these are the ESS conditions of W^* by Definition 2.4.1, the result follows by part (c) of Theorem 5.5.1.

∎

THEOREM 5.5.3. Suppose (p^*, N^*) is a hyperbolic equilibrium of (5.5.1). [That is, no eigenvalues of the linearized dynamic have zero real part.] Then (p^*, N^*) is l.a.s. if and only if (p^*, N^*) is a DDESS of $W(N)$.

PROOF. If (p^*, N^*) is a DDESS then it must satisfy the regularity condition of Theorem 5.4.3 (otherwise there is a zero eigenvalue). Thus (p^*, N^*) is l.a.s.

Conversely, suppose (p^*, N^*) is l.a.s. The proof that (p^*, N^*) is a DDESS follows that of Theorem 5.4.3. First, since there are no zero eigenvalues, the linearization (5.4.1) implies $e_i \cdot W^*p^* < p^* \cdot W^*p^*$ whenever $p_i^* = 0$. Therefore, we may assume $p_i^* > 0$ for all $i = 1, \cdots, n$ and that $p \cdot W^*p^* = 0$ for all $p \in \Delta^n$. Furthermore, from (5.4.1), the linearization $L_{ij} = N_i^* \left(\dfrac{1}{N^*} w_{ij}^* + e_i \cdot W'p^* \right)$ satisfies

$$\langle x, Lx \rangle = \frac{1}{N^*} x \cdot W^*x, \quad \langle x, Lv^* \rangle = 0, \quad \text{and} \quad \langle v^*, Lv^* \rangle = p^* \cdot W'p^*$$

where $v^* = \dfrac{1}{N^*}(N_1^*, \cdots, N_n^*)$ and $x \in X^n$. Thus, the eigenvalues of L are those of W^*

restricted to X^n as well as $p^* \cdot W'p^*$. Since W^* is symmetric, its eigenvalues are negative if and only if $x \cdot W^* x < 0$ for all nonzero $x \in X^n$ (i.e. iff p^* is an ESS of W^*). By part (c) of Theorem 5.5.1, (p^*, N^*) is a DDESS.

∎

The proof of Theorem 5.5.3 is essentially that of Ginzburg (1977). [It is unknown how far the assumption of hyperbolicity can be relaxed.] Both he and Roughgarden (1979) emphasized the equivalence of l.a.s. equilibria of (5.5.1) with the isolated local maxima of ϕ in Theorem 5.5.2. Their biological interpretation suggests a group selection principle is operative here that replaces the principle of maximizing population mean fitness in Theorem 4.1.2. Specifically, the allelic frequencies evolve to maximize population size subject to the equilibrium constraint that the population's growth rate be zero.

To my mind, the above emphasis is misplaced and the more significant equivalence occurs between l.a.s. equilibria and DDESS's (Theorem 5.5.3). From this viewpoint, Theorems 5.5.1 and 5.5.2 exploit mathematical consequences of symmetry for the "payoff" matrices of density-dependent natural selection. [The consequences also have considerable practical importance. For example, Theorem 5.5.1 shows stability can indeed be verified through the separate consideration of the frequency and density dynamic as suggested after Definition 5.2.4.]

Nevertheless, these consequences may inhibit a deeper understanding of selection models when there is a bona fide frequency dependence. [A similar misplaced emphasis was discussed in Section 4.1.] For instance, Roughgarden (1979) devoted much effort generalizing his principle of maximizing population size to coevolutionary models combining interspecific density effects with intraspecific frequency effects. In my opinion, the DDESS concept generalized to two (or more) species by integrating Chapters 3 and 5 holds the promise of unifying his many results.

5.6 Evolutionary Stability in Multi-Species Population-Dynamic Models

This last section of Chapter 5 connects haploid DDESS theory with interspecific population dynamics where each species is monomorphic and the per capita growth rate of species i depends only on the population sizes $N_1, N_2, \cdots, N_n$ of the n different species. That is, the continuous dynamical system has the form

$$\dot{N}_i = N_i F_i (N_1, \cdots, N_n) \tag{5.6.1}$$

where F_i is the (nonlinear) growth rate. For instance, the Lotka-Volterra differential equations (Hofbauer and Sigmund 1988) that model diverse ecosystems involving competition, mutualism and/or predator-prey relationships between species assumes F_i combines a linear function of $N_1, \cdots, N_n$ with an intrinsic growth rate of species i.

There are different ways to place (5.6.1) in the context of evolutionary game theory. Hofbauer and Sigmund (1988) exploit an ingenious equivalence of (5.6.1) to an ESS dynamic of the form (2.5.1) on Δ^{n+1}. A more obvious connection, developed below, is to the DDESS dynamic (5.1.1). The discussion emphasizes the game-theoretic perspective and provides further references for the interested reader.

The above population dynamic can always be written in the form (5.1.1) by setting S_i equal to the ith pure strategy and considering N_i as the number of individuals using strategy S_i. The state (S, N) uniquely determines the size of each species, and conversely, since N_i is equal to the ith component of $NS \in R^n$. Thus $F_1(N_1, \cdots, N_n), \cdots, F_n(N_1, \cdots, N_n)$ determine a unique fitness vector $F(S, N)$ for (5.1.1). [The equivalence between the pure-strategy density-dependent single-haploid-species dynamic and (5.6.1) is not surprising given the fact individuals produce identical offspring in each model.]

For the conventional population dynamic (5.6.1), ecosystem stability at an equilibrium is usually verified by investigating the linearized eigenvalues (i.e. essentially the method used to prove many stability results in this text such as Theorem 5.4.3). However, the fundamental purpose of evolutionary game theory is to provide alternative static conditions, preferably based on biological intuition, that guarantee stability. In particular, the DDESS conditions of Definition 5.4.2, based on the relative frequency distribution of the different species and on the entire ecosystem density, provide such an alternative.

To me, the DDESS approach is biologically superior. However, its mathematical superiority depends on the actual form of the fitness functions F_i. This ambivalence is very clear in Cressman (1988c) where stability for two-species population models is characterized using the frequency/density approach of DDESS theory. However, we soon encounter the same difficulty as in Section 2.6 where non-ESS stable equilibria exist for the pure-strategy dynamic. In such cases, one can argue the interspecific interaction parameters will change in evolutionary time (the Lotka-Volterra example of Cressman (1990) showed evolutionary forces favor behavioral change in both predator and prey species) much like the emergence of super-strategies in Sections 2.4 and 5.4.

AN INTERMISSION

As implied in the Introduction, the end of Chapter 5 marks a clear turning point in the text. Perhaps you will allow me, as a Canadian, to use a game that I am familiar with to describe informally where we have been and where we are going.

The hockey game pits a relatively young home team (evolutionary game theorists) against an older, slightly rusty visiting team (the skeptics). Unexpectedly, only the home team appears for the pre-game warm-up (Chapter 2). It entertains the fans (the readers) with a dazzling display of well-practiced precision drills (existing single species ESS theory). Towards the end of the warm-up (Sections 2.8 to 2.10), some fans grow weary of the more intricate patterns (centre manifold theory).

In the first period (Chapter 3), the crowd sees immediately the versatility (to two-species systems) of the home team's training (the ESS program). The visitors should have at least watched the warm-up. The home team quickly builds an overwhelming lead (to Section 3.5) and is working on a shutout until it becomes over-ambitious (multi-species models) and allows a disputed goal in the final minute (Section 3.6).

This success encourages the visitors to try a classical attack (natural selection) at the start of the second period (Chapter 4). When the younger team easily defends and mounts an effective counter-offensive (Section 4.1), the visitors introduce a modern variation (frequency-dependence). The hockey here is at its most exciting with end-to-end rushes (Sections 4.3 and 4.5) as well as furious checking and body contact at centre ice (Section 4.4). [The latter is the type of hockey that I (used to!) play best.] The visitors mildly protest when the broadcast announcer (the author) declares the home team won the period. In the final period (Chapter 5), the visitors probe a possible weekness (Section 5.2 versus Section 5.4 DDESS definitions) in their opponent's training. The home team again defends vigorously (Sections 5.3 and 5.4) — managing to score the final goal (Section 5.5) on an inspired variation (density-dependent natural selection) of a second period play.

Both teams and the fans remember this hockey game clearly. It came to represent one of the few games in the history of the sport (evolutionary biology) that marked a clear shift in the accepted style of play (shift in theories). The fans demanded further exhibition games (Chapter 6 examples). In these, the home team gained confidence in applying its style to new situations (new models) and the visitors kept it from becoming complacent.

In hockey, intermissions occur between periods to clear the ice. [Perhaps, there should have been an intermission between each chapter.] They also permit both teams to rest. I will have accomplished my goal in this intermission if it has allowed the fans time to clear their minds and rest up for the action in Chapter 6.

6. EVOLUTIONARILY STABLE SETS AND CONTESTANT INFORMATION

The preceding four chapters have each analyzed a general theoretical model in evolutionary biology. The limited examples of particular evolutionary games played a secondary role to the theory. That is, their main purpose was to illustrate the theory at hand.

This final chapter's approach is different. Here, the examples take centre stage. Each evolutionary game is intended to examine an area in evolutionary biology that has clear game-theoretic overtones (e.g. the effect of territorial ownership on an individual's behavior). The danger to this reverse emphasis is that the chapter may appear to be an eclectic collection of examples that lacks a central focus. To counteract this possibility, I will continue to consider both static and dynamic aspects of these games as well as connect them to previous theory. Moreover, there are two common threads that run through most of the sections.

The first is that the evolutionary games do not necessarily have a single solution strategy, such as an ESS, that is l.a.s. — rather there are many solutions that all exhibit a neutral stability. In particular, the neutrally-stable evolutionarily stable sets (ES Sets) introduced in Section 6.2 appear in many of the examples. As we will see (Theorem 6.3.2), an ES Set is a set of noninvadable strategies just as an ESS is a single strategy that is noninvadable.

The second common feature is that most examples involve individuals who have more information concerning the evolutionary game beyond the knowledge of the payoff matrix. For instance, an individual may know inherently some aspect of the environment (Section 6.4), of the opponent's relative status (Section 6.6), or of the opponent's choice of strategy in previous confrontations (Section 6.7). This contrasts markedly with all examples considered so far where only the payoffs from a single contest are known.

Neither of these features are meant to suggest evolutionary game theory developed in Chapters 2 to 5 is now irrelevant. Indeed, for me, Chapter 6 deepens the heuristic appreciation of the game-theoretic principles emphasized until now and points to the future maturity of the theory of evolutionary games.

6.1 A Mixed-Strategy Hawk-Dove Game

ES Sets often occur in game-theoretic models when the biologist and/or modeller fails to recognize that some of the observed individual phenotypes of a particular species are actually mixed strategies. For example, suppose an observer has determined this species has exactly three types of individuals and has also determined the numerical payoffs for each of the nine possible pairs of individual interactions. Furthermore, suppose that, unknown to the observer, all three phenotypes are strategies S_1, S_2 and S_3 in Δ^2 of the hawk-dove game from Section

2.3 that has payoff matrix $A = \begin{bmatrix} -1 & 2 \\ 0 & 1 \end{bmatrix}$ (i.e. $V = 2$ and $C = 4$). We will assume that the

strategies S_1 and S_2 are the pure strategies H and D respectively while S_3 is the mixed strategy $(1/2, 1/2) \in \Delta^2$ that plays H half the time.

According to our observer, the 3×3 payoff matrix B must be

$$
\begin{array}{c}
\quad\quad S_1 \quad\ S_2 \quad\ S_3 \\
\begin{array}{c} S_1 \\ S_2 \\ S_3 \end{array}
\begin{bmatrix}
-1 & 2 & 1/2 \\
0 & 1 & 1/2 \\
-1/2 & 3/2 & 1/2
\end{bmatrix}.
\end{array}
\tag{6.1.1}
$$

[For instance, the payoff to S_1 against S_3 is $e_1 \cdot A(1/2, 1/2) = e_1 \cdot (1/2, 1/2) = 1/2.$]
As mentioned after the payoff matrix (2.4.7), $Bp = 1/2(1, 1, 1)$ for any frequency vector, p,
in the set $L = \{ p = (p_1, p_1, 1 - 2p_1) \,|\, 0 \le p_1 \le 1/2 \}$. Thus any such $p \in L$ is an
equilibrium of the continuous haploid dynamic

$$
\dot{p}_i = p_i(e_i - p) \cdot Bp \qquad i = 1, 2, 3
\tag{6.1.2}
$$

that can be regarded as a pure-strategy model on Δ^3. In particular, there is no ESS for the matrix B.

On the other hand, any initial polymorphic population will evolve under (6.1.2) to a point on L (page 132, Hofbauer and Sigmund 1988) as indicated by the flow diagram of Figure 6.1.

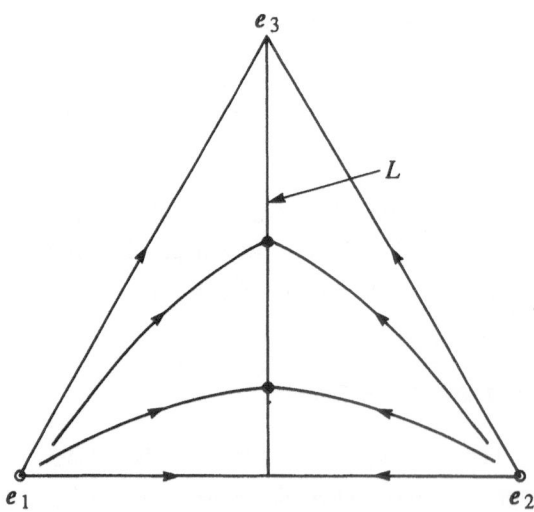

Figure 6.1. *Flow diagram for the dynamic (6.1.2)*

All points on L are neutrally stable equilibria that correspond to the ESS of the underlying hawk-dove game.

To prove this, one can show $V(p) = -(p_1 - p_2)^2$ is a strict Liapunov function on the interior of Δ^3 since

(i) $V(p) \leq 0$ with equality only for points in L, and

(ii) $\dot{V}(p) = (p_1 - p_2)^2(p_1 + p_2 - (p_1 - p_2)^2) > 0$ unless $p \in L$.

Since all points in L correspond to a population mean strategy $S_0^* = (1/2,\ 1/2)$ in the underlying hawk-dove game, this dynamic behavior is no surprise to the modeller who does recognize that the 2×2 matrix A is the primary payoff matrix. Figure 6.1 then verifies that the mean strategy evolves (monotonically) to the interior ESS S_0^* of A. [The analysis also agrees with the discussion in Section 2.8 (B) in that L is the centre manifold for the linearization of (6.1.2) about an equilibrium p in the interior of Δ^3.]

At the same time, a less-perceptive observer should have at least noticed that the equilibrium points $p \in L$ are neutrally stable but not l.a.s. because a small difference in the initial mixture of the three phenotypes produces a different limiting frequency vector in L. These observations cannot be explained through the ESS theory developed so far as it applies to the payoff matrix B. However, an intuitive explanation is possible by considering the ES Sets of B (see comment after Theorem 6.2.5 below).

6.2   The Static Characterization of an ES Set

This section and the next give the static and dynamic characterizations, respectively, of general ES Sets. I have attempted to keep these theoretical sections short — limiting them to the material essential for an understanding of the examples in the rest of the chapter.

The dynamic result from the previous section (i.e. L is a set of neutrally stable equilibria) is not at odds with ESS theory developed in Chapter 2 for the matrix (6.1.1). In fact, we might have predicted this result since points in L are almost ESS's according to Definition 2.4.1. [Specifically, all points in L satisfy the symmetric Nash equilibrium condition (2.4.1) and only fail to be ESS's since the stability condition (2.4.1) is not a strict inequality with respect to other points in L.] Such considerations led Thomas (1985a) to introduce ES Sets.

DEFINITION 6.2.1. Let A be an $m \times m$ payoff matrix. A *(symmetric Nash) equilibrium strategy* $S_0 \in \Delta^m$ satisfies, for all $S \in \Delta^m$,

$$S \cdot AS_0 \leq S_0 \cdot AS_0. \tag{6.2.1}$$

An *evolutionarily stable set (ES Set)* L is a set of equilibrium strategies that satisfies

$$S \cdot AS < S_0 \cdot AS \tag{6.2.2}$$

whenever $S_0 \in L$, $S \notin L$ and there is equality in (6.2.1).

Comparison with Definition 2.4.1 shows (2.4.1) and (6.2.1) are identical whereas (2.4.2) need only hold for strategies outside the ES Set. For example,

$$L = \{p = (p_1, p_2, p_3) \in \Delta^3 \,|\, p_1 = p_2\}$$

is an ES Set of matrix (6.1.1) since, for all $p_0 \in L$,

(i) $Bp_0 = \frac{1}{2}(1, 1, 1)$ and

(ii) $p \cdot Bp - p_0 \cdot Bp = -\frac{1}{2}(p_1 - p_2)^2 < 0$ if $p \notin L$.

Furthermore, L is the only ES Set by Theorem 6.2.5 below. It is probably the simplest example of an interesting ES Set. [ES Sets are of little interest for two-strategy games. If a 2×2 matrix A is not selectively neutral, the only ES Sets consist of sets of ESS's. If A is selectively neutral, the entire strategy simplex is the only ES Set.]

It is easy to see that S^* is an ESS for a general $m \times m$ payoff matrix A if and only if the singleton set, $\{S^*\}$, is an ES Set. Also clear is that any set containing only ESS's is an ES Set and that any finite union of ES Sets is an ES Set. Not quite as obvious is that the construction of the ES Set L in Section 6.1 generalizes to any mixed-strategy game. Indeed, we have

THEOREM 6.2.2. (Thomas 1985b) Suppose an ESS, S_0^*, of an $m \times m$ payoff matrix A is some convex combination, say $S_0^* = \Sigma p_i^* S_i$, of the (mixed) strategies $S_1, S_2, \cdots, S_n$. Then $L = \{p \in \Delta^n \,|\, S(p) = S_0^*\}$ is an ES Set for the $n \times n$ payoff matrix B with entries $B_{ij} = S_i \cdot AS_j$. [Recall that $S(p)$ is the mean strategy $\Sigma p_i S_i \in \Delta^m$.]

PROOF. Since $p \cdot Bq = \Sigma p_i S_i \cdot A \Sigma p_i S_i = S(p) \cdot AS(q)$,

$$p \cdot Bp_0 = S(p) \cdot AS_0^* \le S_0^* \cdot AS_0^* = p_0 \cdot Ap_0 \qquad (6.2.3)$$

for all $p_0 \in L$ by (2.4.1). Also, if $p \notin L$, then $S(p) \ne S_0^*$ so that

$$p \cdot Bp = S(p) \cdot AS(p) < S_0^* \cdot AS(p) = p_0 \cdot Bp$$

if there is equality in (6.2.3).

∎

Thus, ESS's of one payoff matrix induce ES Sets for corresponding mixed-strategy payoff matrices. [Examples later in this chapter show this is not the only way ES Sets occur.] After the dynamic characterization of ES Sets is completed in Section 6.3, our observer in Section 6.1 will be able to predict the evolutionary outcome solely on the basis of matrix (6.1.1). In the meantime, the geometric characterization of ES Sets is described in the remainder of this section. It rests on the fact that knowing certain elements in the ES Set can provide the complete description in much the same way that knowing a $p \in L$ in Section 6.1 determines all of L as $\{p' \in \Delta^3 \,|\, S(p') = S(p)\}$.

From Definition 6.2.1, it is clear that an ES Set containing S must also contain any S' that is "fitness neutral" with respect to S in the following sense

$$S' \cdot AS = S \cdot AS \quad \text{and} \quad S \cdot AS' = S' \cdot AS'. \tag{6.2.4}$$

In particular, this ES Set must contain all strategies that are in L_S as given by

DEFINITION 6.2.3. Suppose $S \in \Delta^m$ is a strategy in some ES Set L. Define L_S as $\{S' \in \Delta^m \,|\, \text{supp}\, S' \subseteq \text{supp}\, S \quad \text{and} \quad e_k \cdot AS' = e_l \cdot AS' \text{ for all } k, l \in \text{supp}\, S\}$. S has *maximal support in* L if no other $S' \in L$ has support properly containing $\text{supp}\, S$.

Intuitively, L_S is the set of strategies that are fitness neutral to S and whose support is contained in that of S. Furthermore, if S has maximal support, then L_S should contain all those strategies that are fitness neutral with respect to S. The following theorem formalizes these ideas.

THEOREM 6.2.4. Suppose S is in the ES Set L. Then
(a) $S \in L_S = \{S' \in L \,|\, \text{supp}\, S' \subseteq \text{supp}\, S\}$
(b) L_S is an ES Set if and only if S has maximal support in L.

PROOF. (a) To show $S \in L_S$, we need to show $e_k \cdot AS = e_l \cdot AS$ for all $k, l \in \text{supp}\, S$. Suppose the maximum value of $e_k \cdot AS$ for all $k \in \text{supp}\, S$ occurs when $k = k_0$. If $e_k \cdot AS < e_{k_0} \cdot AS$ for some $k \in \text{supp}\, S$, then $e_{k_0} \cdot AS > S \cdot AS$ and this contradicts (6.2.1). Therefore $S \in L_S$. Furthermore, if $S' \in L_S$, then we have $S' \cdot AS = S \cdot AS$ and $S \cdot AS' = S' \cdot AS'$. Thus, from (6.2.4), $S' \in L$.

Conversely, assume that $S' \in L$ and $\text{supp}\, S' \subseteq \text{supp}\, S$. If $e_k \cdot AS' \neq e_l \cdot AS'$ for all $k, l \in \text{supp}\, S$, let $e_{k_0} \cdot AS'$ be the smallest and $e_{l_0} \cdot AS'$ be the largest. By the same reasoning as the preceding paragraph, $S' \cdot AS' = e_{l_0} \cdot AS'$. Thus $S \cdot AS' < S' \cdot AS'$ since $k_0 \in \text{supp}\, S$. Since $S' \cdot AS = S \cdot AS$ and S is in an ES Set, this last inequality contradicts (6.2.2) and so $S' \in L_S$.

The proof of (a) also shows that elements in L_S are indeed those strategies with support in $\text{supp}\, S$ that are fitness neutral with respect to S (see (6.2.4)).

(b) Assume L_S is an ES Set. If $S' \in L$ and $\text{supp}\, S \subseteq \text{supp}\, S'$, then $S \in L_{S'}$ by part (a). In particular, S and S' are fitness neutral and so S' must be in any ES Set that contains S. Thus, $S' \in L_S$ and so $\text{supp}\, S' = \text{supp}\, S$. That is, S has maximal support in L.

Conversely, assume S has maximal support in an ES Set L. If L_S is not an ES Set, then there must be (by Definition 6.2.1) an S' in L but not in L_S and an $S'' \in L_S$ such that

$$S' \cdot AS'' = S'' \cdot AS'' \quad \text{and} \quad S'' \cdot AS' \leq S' \cdot AS'. \tag{6.2.5}$$

It is easy to verify that S' replaced by $1/2(S' + S'')$ also satisfies (6.2.5). Furthermore, $1/2(S' + S'')$ is in L but not in L_S. If $\text{supp}\, S'' = \text{supp}\, S$, then $\text{supp}\, 1/2(S' + S'')$ properly contains supp S and we have contradicted the fact that S has maximal support. If supp $S'' \neq$ supp S, a contradiction is reached by considering the subgame of A based on the pure strategies in supp $S'' \cup$ supp S'. [Alternatively, the local form of an ES Set (Definition 6.3.4 in the following section) can be used to reach a contradiction.] Thus L_S is an ES Set.

∎

Theorem 6.2.4 is the generalization of Haigh's (1975) observation that the support of one ESS cannot be contained in the support of another. For any ES Set L, the supports of elements in L form a partially ordered (under inclusion) set of subsets of $\{1, 2, \cdots, m\}$. To describe L geometrically, pick one $S \in L$ with the appropriate support to represent each of the finite number of maximal supports. Then L is the finite union of these L_S. [It does not matter which S is chosen since L_S depends only on the support of S.] By Definition 6.2.3, each L_S is the intersection of Δ^m with an $(m - d)$ dimensional hyperplane (d is the number of pure strategies in the support of S). That is, we have proven

THEOREM 6.2.5. (Cressman 1992) Any ES Set L is the finite union of disjoint, closed, connected linear submanifolds. These submanifolds are all ES Sets of the form L_S for some subset of equilibrium strategies in L that have maximal support. Furthermore, if $L_S \neq L_{S'}$, then supp S cannot be contained in supp S'.

One immediate consequence of this theorem is especially useful in Section 6.1. If L_S is an ES Set and supp $S = \{1, 2, \cdots, m\}$, then L_S is the only ES Set. In particular, the unique ES Set of matrix (6.1.1) is L_S where $S = (1/4, 1/4, 1/2)$.

6.3 The Dynamic Characterization of an ES Set

In Figure 6.1, the continuous pure-strategy haploid dynamic (6.1.2) based on the 3×3 payoff matrix (6.1.1) has the ES Set L as the global attractor for any initial polymorphism. Indeed, given an initial mixture of pure-strategists, the population evolves to a particular equilibrium strategy in L. Qualitatively, the same dynamic result is true locally for general ES Sets under the mixed-strategy dynamic (Theorem 6.3.2 below).

As in Section 2.5, the mixed-strategy dynamic

$$\dot{p}_i = p_i(S_i - S) \cdot AS \qquad (6.3.1)$$

induces a dynamic of the population's mean strategy $S = \Sigma p_i S_i \in \Delta^m$ that is a convex combination of the phenotypes $S_1, \cdots, S_n$. Then any ES Set L satisfies the following two properties by Theorem 6.3.2.

(i)   If $S_0 = \Sigma p_i^* S_i \in L$, then p^* is an equilibrium of (6.3.1).

(ii) L is a *local attractor* (i.e. if S is sufficiently close to S_0, then S evolves to the set L) of the mean strategy evolution determined by (6.3.1).

Moreover, these properties completely characterize ES Sets through a strong stability concept.

DEFINITION 6.3.1. A set L in Δ^m is called *strongly stable* if, whenever $S_0 \in L$ is a convex combination of $\{S_1, S_2, \cdots, S_n\}$, S_0 is an equilibrium of, and the connected component of L that contains S_0 is a local attractor of, the mean strategy evolution determined by (6.3.1).

THEOREM 6.3.2. L is strongly stable if and only if L is an ES Set.

The above discussion generalizes Definition 2.6.1 and Theorem 2.6.2 to sets of equilibrium strategies. The technically demanding proof of Theorem 6.3.2 (Cressman 1992) becomes substantially easier (and more understandable!) if L contains a strategy S_0 in the interior of Δ^m. The following "proof" makes this simplifying assumption.

PROOF. Suppose $S_0 \in L$ is in the interior of Δ^m. Assume L is strongly stable. Since S_0 is an equilibrium of (6.3.1) applied to the pure-strategy dynamic, $e_i \cdot AS_0 = S_0 \cdot AS_0$ for $i = 1, 2, \cdots, m$. Thus S_0 is an equilibrium strategy by (6.2.1) and L must contain $E = \{S \in \Delta^m \mid AS = \lambda 1 \text{ for some } \lambda \in R\}$ since this set of equilibria for (6.3.1) forms a connected linear submanifold through S_0. In fact, any $S \in L$ that is in the interior of Δ^m must be in E (any other S is not an equilibrium) and vice versa. Thus, if $S_1 \notin L$ and S is in the interior of Δ^m, then S_0 is the only element of L that lies on the line segment joining S_1 to S_0. The one-dimensional dynamic (6.3.1) with respect to the pair $\{S_0, S_1\}$ of mixed strategies is

$$\dot{p}_0 = p_0(S_0 - S) \cdot AS = p_0(1 - p_0)^2(S_0 - S_1) \cdot AS_1 \qquad (6.3.2)$$

since $S = p_0 S_0 + (1 - p_0)S_1$. Since L is strongly stable, p_0 must locally evolve to 1. In particular,

$$S_1 \cdot AS_1 < S_0 \cdot AS_1. \qquad (6.3.3)$$

By linear algebra, (6.3.3) can be extended to any $S_0 \in E$ and any $S_1 \notin E$. Thus, by (6.2.2), E is an ES Set. It remains to show that $L = E$. If there is an $S_1 \in L$ that is not in E, then (6.3.3) implies the connected component of L that contains S_1 is not a local attractor of (6.3.2) (a contradiction of strong stability).

The converse follows the proof of Theorem 2.5.2. Assume L is an ES Set. By Theorem 6.2.5, $L = L_{S_0}$ and so all points of L are equilibria of (6.3.1). Furthermore, if $S^* \in L$ is some convex combination of $\{S_1, S_2, \cdots, S_n\}$, say $S^* = \Sigma p_i^* S_i$, then $V(p) = \Pi p_i^{p_i^*}$ is a strict Liapunov function of (6.3.1) as in the proof of Theorem 2.5.2. That is, by (6.2.2),

$$\dot{V}(p) = V(p)(S^* - S) \cdot AS \geq 0$$

with equality iff S is an equilibrium strategy in L. If p^* in the definition of $V(p)$ is replaced by a p^∞ in the limit set of the dynamic, then $S(p^\infty) \in L$ and (6.3.1) evolves to p^∞. That is, L is a local attractor. ∎

REMARK 6.3.3. The above proof shows that an ES Set is a global attractor if an $S_0 \in L$ is in the interior of Δ^m and S_0 is a convex combination of the phenotypes initially present. If L does not contain a strategy in the interior of Δ^m, the Liapunov function used in the above proof remains valid locally as can be seen by the following definition of ES Sets (Thomas, 1985b). [Although this definition, which is equivalent to Definition 6.2.1, is not emphasized in this text, it is perhaps the one that is the more appealing to biological intuition in the same way that Definition 2.4.2 compares to Definition 2.4.1.]

DEFINITION 6.3.4. L is an *ES Set* for the $m \times m$ payoff matrix A if, for every $S_0 \in L$, there is a neighborhood U of S_0 in Δ^m such that

$$S_0 \cdot AS \geq S \cdot AS$$

for all $S \in U$ with equality if and only if $S \in L$.

REMARK 6.3.5. Through the above local definition, Thomas (1985c) also extended the notion of ES Sets to the simplex Δ^n of single-locus allele frequencies p for the diploid population of Chapter 4. The nonlinearity of the mean strategy map $S(p) = \Sigma p_i p_j S_{ij}$ from Δ^n to Δ^m meant ES Sets were defined in terms of nonlinear fitness functions on Δ^n. Thomas was then able to show Theorem 6.2.2 remains valid (with B replaced by these fitness functions). That is, an ESS S^* in the strategy simplex Δ^m produces an ES Set in Δ^n. However, just as we failed in Section 4.4 (e.g. Theorem 4.4.1), Thomas fell just short of proving S^* was l.a.s. (his additional requirements to prove dynamic stability translate into conditions on the centre manifold of Section 4.4 (B)). Nevertheless, I encourage readers interested in genetic ESS theory to gain a better appreciation of Thomas' perspective.

6.4. The Hawk-Dove Game with Varying Resource

The description of those ES Sets that correspond to an ESS of a mixed-strategy game (such as the ES Set of Section 6.1) appeals to biological intuition through Theorems 6.2.2 and 6.3.2. However, in general, the biological description of ES Sets given in Theorem 6.2.5 may be quite difficult. The biological model developed in this section illustrates one such difficulty. It is also the first model in this text where strategies are based on information other than the payoffs of a single contest.

Specifically, suppose individuals engage in pairwise contests over either a good resource ($V = 10$) or a poor one ($V = 5$) and that the cost of fighting is the same ($C = 20$) in either case. Furthermore, assume that both contestants know the value of the resource under dispute and that 60% of the time the resource is good (the other 40% it is poor). There are four pure strategies in this game; labelled e_1, e_2, e_3, and e_4.

(i)   Play H in all contests.

(ii)   Play H in contests when $V = 10$ and D when $V = 5$.

(iii)   Play D when $V = 10$ and H when $V = 5$.

(iv)   Play D in all contests.

In a single contest, payoffs are given by the two payoff matrices

$$\begin{array}{cc} & \begin{array}{cc} H & D \end{array} \\ \begin{array}{c} H \\ D \end{array} & \begin{bmatrix} -5 & 10 \\ 0 & 5 \end{bmatrix} \end{array} \quad \text{and} \quad \begin{array}{cc} & \begin{array}{cc} H & D \end{array} \\ \begin{array}{c} H \\ D \end{array} & \begin{bmatrix} -7.5 & 5 \\ 0 & 2.5 \end{bmatrix} \end{array} \tag{6.4.1}$$

when the resource is good or poor respectively. The expected payoff to the second pure strategist in conflict with the third is $(0.6)(10) + (0.4)(0) = 6$ since 60% of the time it plays H against D (i.e. when $V = 10$) to obtain payoff 10 and the other 40% it plays D against H (i.e. when $V = 5$) to obtain payoff 0. Thus $A_{23} = 6$ in the following 4×4 payoff matrix A that represents this game.

$$A = \begin{bmatrix} -6 & -1 & 3 & 8 \\ -3 & -2 & 6 & 7 \\ -3 & 2 & 0 & 5 \\ 0 & 1 & 3 & 4 \end{bmatrix}. \tag{6.4.2}$$

LEMMA 6.4.1. $L \equiv \{S \in \Delta^4 \mid S = (x, \ 1/2 - x, \ 1/4 - x, \ 1/4 + x), \ 0 \le x \le 1/4\}$ is the only ES Set of A.

PROOF. Since $AS = 9/4(1, \ 1, \ 1, \ 1)$ for any $S \in L$, L is a set of equilibrium strategies. Secondly, A is negative semi-definite on X^4 because $z = -(x + y + w)$ and

$$\begin{aligned} (x, \ y, \ z, \ w) \cdot A(x, \ y, \ z, \ w) &= -6x^2 - 4xy + 8xw - 2y^2 + 8yz + 8yw + 8zw + 4w^2 \\ &= -2(3x^2 + 5y^2 + 2w^2 + 6xy + 4yw) \\ &= -6(x + y)^2 - 4(y + w)^2 \\ &\le 0 \end{aligned}$$

with equality iff $y = -x$, $w = x$ and $z = -x$. Therefore, if $S \in L$, then $(S' - S) \cdot AS' = (S' - S) \cdot A(S' - S) < 0$ for all $S' \in \Delta^4$ unless $S' - S = (x, \ -x, \ -x, \ x)$ (i.e. unless $S' \in L$). By (6.2.2), L is an ES Set. L is the only ES Set since $(1/8, \ 3/8, \ 1/8, \ 3/8) \in L$ is in the interior of Δ^4. ∎

By Remark 6.3.3, L is the global attractor of the mean-strategy evolution under the pure-strategy dynamic for any initial polymorphism. What is the intuitive description of L? Some properties of L listed below are biologically reasonable and others are more questionable.

(i)   The population's mean strategy does evolve to one that makes some use of the information on the resource's value (i.e. no $S \in L$ has support contained in $\{1, \ 4\}$). [Reasonable]

(ii)   Any $S \in L$ includes some individuals that make no use of this extra information. [Questionable]

(iii) There are always more individuals using e_2 than e_3 in the limiting population. [Reasonable since H should be used when the resource is good if it is used when the resource is poor.]

(iv)   There are some $S \in L$ that have individuals using e_3.   [Questionable for the same reason as given in (iii).]

The above list alternates between intuitive and counterintuitive qualitative results for this simple game-theoretic model that incorporates some information on the environment. It turns out L does have an obvious biological interpretation that has been hidden through the use of the *normal form* (i.e. payoff matrix (6.4.2)) for this model. I feel the extensive form of such games (a topic ignored by most researchers in evolutionary game theory) and especially the concept of a direct ESS is more appropriate here. [The discussion in the next section, where extensive games and direct ESS's are introduced, should convince the reader of this as well.]

6.5 ES Sets for Games in Extensive Form

The complete formal definition of general evolutionary games in extensive form, introduced by Selten (1983), is technically demanding. I have attempted to avoid these formalities by describing the theory as it applies to the hawk-dove game with varying resource. It is then hoped that these ideas can be transferred by the reader to the other extensive models considered in the next two sections. For the complete description of extensive evolutionary games see Selten (1983) or Cressman (1992) where, for instance, definitions are provided for the terms given here it italics.

The model of Section 6.4 is represented in extensive form by a *game tree,* Figure 6.5, that is divided into two branches at the *starting point* where a *random decision* is made to follow the good resource branch or the poor one with probabilities 0.6 and 0.4 respectively. On each branch, a single hawk-dove game is played between two players who do not know the other's decision. The game tree emphasizes the sequential nature of the game (first the random decision, then player 1's decision and finally player 2's decision) and indicates what information each decision is based on. For instance, player 2 cannot base its decision on player 1's *choice* since player 2's two *decision points* on each half of the tree are enclosed in an *information set* given by the dashed curve in Figure 6.5. That is, player 2 must make the same choice at both decision points.

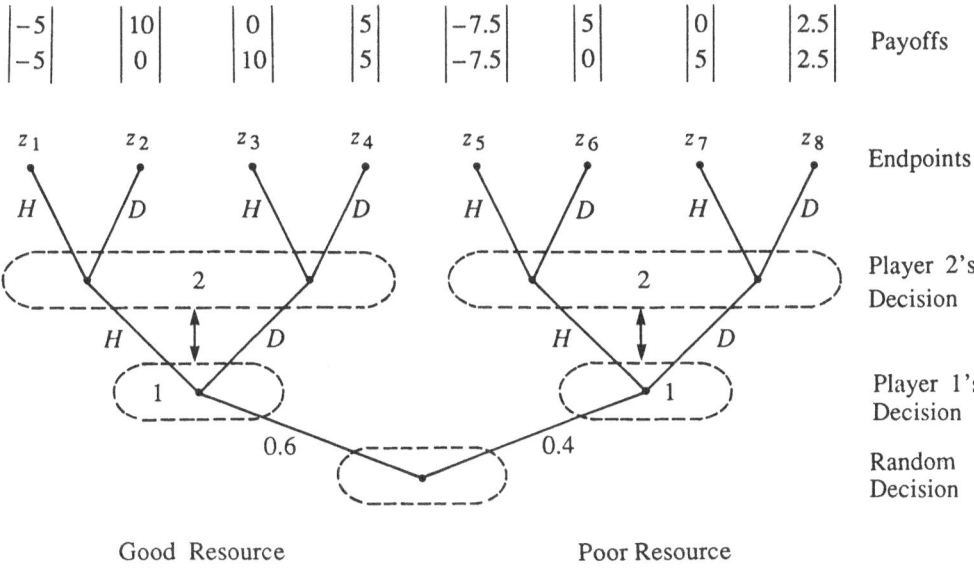

Figure 6.5. *The hawk-dove game with varying resource in extensive form*

Extensive evolutionary games (for a single species) also include a *symmetry* ($\leftrightarrow$ in Figure 6.5) between the decision points of player 1 and of player 2. In our example, the symmetry is between their decisions in contests over resources of the same value. There are eight possible *plays* of the game corresponding to the *endpoints* z_1, z_2, $\cdots$, z_8. For each play, there are *payoffs* $\begin{vmatrix} a \\ b \end{vmatrix}$ for the two players (a for player 1 and b for player 2) given by the appropriate entries of (6.4.1). A player has four *pure strategies* as listed previously in Section 6.4. The *symmetric normal form* is then a 4×4 matrix A whose entries give the payoff of one pure strategy against another. For example, a contest involving e_2 against e_3 has endpoints either z_2 or z_7 and expected payoffs $(0.6)\begin{vmatrix} 10 \\ 0 \end{vmatrix} + (0.4)\begin{vmatrix} 0 \\ 5 \end{vmatrix} = \begin{vmatrix} 6 \\ 2 \end{vmatrix}$ that match the A_{23} and A_{32} entries of (6.4.2). [That is, A is given by (6.4.2).]

A *behavior strategy* for these *symmetric extensive games* assigns a *local strategy* at each information set of one of the players. This local strategy is a mixed strategy (i.e. frequency distribution) over the choices at each decision point in the information set. It is essential to realize that behavior strategies are different than strategies for games in normal form. For instance, the behavior strategy b^* of player 1 that plays H with probability $1/2$ or $1/4$ respectively in contests involving a good or poor resource respectively is not a mixed strategy of the game (6.4.2).

In fact, any mixed strategy in the ES Set for A, given in Lemma 6.4.1 by

$$L = \{S \in \Delta^4 | S = (x,\ 1/2 - x,\ 1/4 - x,\ 1/4 + x) \text{ for some } 0 \leq x \leq 1/4\}, \quad (6.5.1)$$

corresponds to b^* since any such strategy S plays H over good resources with probability $x + (1/2 - x) = 1/2$ and with probability $x + (1/4 - x) = 1/4$ over poor ones. Moreover, b^* can be determined quite easily since its component local strategies are simply the respective ESS's of the two payoff matrices in (6.4.1). This is the biological intuition hidden by the analysis of the game presented in Section 6.4.

To repeat, the ES Set of Section 6.4 is the behavior strategy that chooses the ESS strategy for the hawk-dove game appropriate to each half of Figure 6.5. This behavior strategy is an example of a *direct ESS*. These can be defined for general symmetric extensive games through static fitness comparisons, similar to Definition 2.4.1, between individual behavior strategies. Cressman (1992) proved that all direct ESS's correspond to ES Sets in symmetric normal form but not conversely. The complete biological description of arbitrary ES Sets in extensive games remains an open problem.

REMARK 6.5.1. I find the conventional terminology "behavior strategy" and "direct ESS" somewhat confusing since neither phrase suggests the local decision-making process inherent in this approach. Other possibilities such as "local ESS" or "conditional strategy" may be more descriptive but are also ambiguous because they could easily refer to other contexts. As a compromise, the above conventional terminology will be maintained for the time being.

REMARK 6.5.2. The description of an individual's behavior in an extensive game as a mixed strategy in normal form is confusing since many mixed strategies describe the same individual behavior (e.g. all individuals in the set L of (6.5.1) use behavior strategy b^*) unless each local strategy is pure. On the other hand, to analyze the mean-strategy dynamic of a polymorphic population consisting of pure strategists, we must know the mixed mean strategy and not just the mean behavior strategy. [That is, the mean behavior strategy of a population of pure-strategists does not fully determine the frequency of different phenotypes in much the same way that the mean strategy in a population of mixed-strategists in Chapter 2 is not a full description of phenotypic frequency.] However, the mean-strategy dynamic does induce an evolution on the mean behavior strategy. With suitable extensions of strong stability and l.a.s. for behavior strategies, one can translate Theorem 6.3.2 to the following result.

THEOREM 6.5.3. If a direct ESS b^* is a convex combination of the initial behavior strategies, then b^* is a l.a.s. equilibrium for the evolution of direct strategies induced by the dynamic (6.3.1). Furthermore, b^* is strongly stable if and only if b^* is a direct ESS.

6.6 The Owner-Intruder Game

In Section 6.4, individuals based their strategy on environmental information (i.e. whether the resource was good or poor). On the other hand, information concerning the opponent that is readily apparent to both individuals, such as relative size or age, may also be used by biological species without violating our basic assumption that an individual's strategy is not a rational decision process. For instance, a population may evolve to a situation where the larger opponent is the aggressor in any pairwise contest. In the literature, such strategies are said to be based on *informational asymmetry* since opponents are not on an equal footing. [By contrast, strategies in Section 6.4 do not depend on comparisons between opponents.]

One of the best-known biological examples to include informational asymmetry is the model of territorial conflict between the owner of the territory (or resource) and an intruder. In this example, both contestants know who is the owner and who is the intruder. [This knowledge is the informational asymmetry.]

To simplify the analysis, we will assume that both owner and intruder place the same value on the territory and that, if a fight develops between them, each has an equal chance of winning and equal costs of fighting. The winner becomes the owner of the territory. Although a more realistic assumption may be that the owner has an advantage in fighting effectiveness due to familiarity with the territory, we will see that game theory does predict an intuitive result even in our simplified model.

A particular territorial conflict will be modelled by the hawk-dove game with payoff matrix

$$\begin{bmatrix} -4 & 8 \\ 4 & 4 \end{bmatrix}. \tag{6.6.1}$$

One heuristic argument would suggest that the population's mean strategy should evolve to the ESS of this matrix, $S_0^* = (1/2, 1/2)$, for both owners and intruders because payoffs are independent of an individual's ownership position. This will be examined from the following perspective that assumes all individuals belong to the same haploid species.

(A) The Single-Species Extensive-Game Perspective

Here, an individual has four possible pure strategies.

(i) Play H if owner or intruder.
(ii) Play H if owner and D if intruder.
(iii) Play D if owner and H if intruder.
(iv) Play D if owner or intruder.

The extensive game approach from Section 6.5 requires a symmetry in the game tree (Figure 6.6.1) between player 1's and player 2's decisions. Since each contest involves one owner

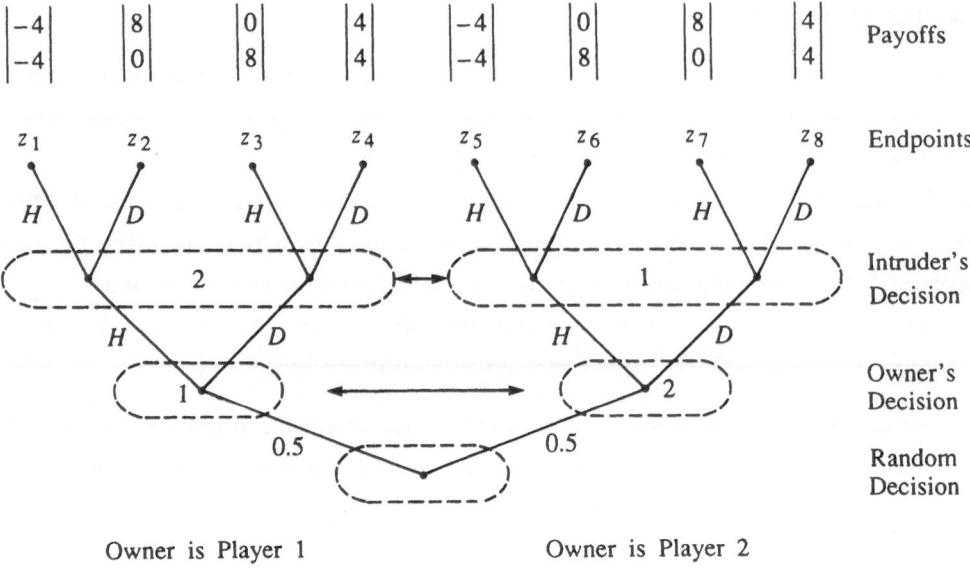

$$\begin{vmatrix} -4 \\ -4 \end{vmatrix} \quad \begin{vmatrix} 8 \\ 0 \end{vmatrix} \quad \begin{vmatrix} 0 \\ 8 \end{vmatrix} \quad \begin{vmatrix} 4 \\ 4 \end{vmatrix} \quad \begin{vmatrix} -4 \\ -4 \end{vmatrix} \quad \begin{vmatrix} 0 \\ 8 \end{vmatrix} \quad \begin{vmatrix} 8 \\ 0 \end{vmatrix} \quad \begin{vmatrix} 4 \\ 4 \end{vmatrix} \qquad \text{Payoffs}$$

Figure 6.6.1. *The owner-intruder game in extensive form*

and one intruder, a natural symmetry (Selten 1983) is between player 1's decision as an owner and player 2's as an owner. To accomplish this, the extensive game in Figure 6.6.1 assumes there is an initial random decision that chooses which player is the owner with equal probability. The next decision is the owner's (between H and D) and finally the intruder chooses a strategy without knowing the owner's choice.

If owners and intruders both play the mixed local strategy $S_0^* = (1/2, 1/2)$ at each decision point, then the game reaches any one of the eight endpoints with equal probability and so the payoffs are

$$1/8 \left(\begin{vmatrix} -4 \\ -4 \end{vmatrix} + \begin{vmatrix} 8 \\ 0 \end{vmatrix} + \begin{vmatrix} 0 \\ 8 \end{vmatrix} + \begin{vmatrix} 4 \\ 4 \end{vmatrix} + \begin{vmatrix} -4 \\ -4 \end{vmatrix} + \begin{vmatrix} 0 \\ 8 \end{vmatrix} + \begin{vmatrix} 8 \\ 0 \end{vmatrix} + \begin{vmatrix} 4 \\ 4 \end{vmatrix} \right) = \begin{vmatrix} 2 \\ 2 \end{vmatrix} .$$

That is, in this case, both players receive payoff 2. Furthermore, any mixed strategy in the set $L = \{(1/4 + x, \ 1/4 - x, \ 1/4 - x, \ 1/4 + x) \, | \, 0 \le x \le 1/4\}$ corresponds to S_0^* since these individuals all play H with probability $1/2$ if owner or intruder.

To determine if L is an ES Set (i.e. if S_0^* is a direct ESS) we need the 4×4 payoff matrix A (Cressman 1992) given in normal form by

$$A = \begin{bmatrix} -4 & 2 & 2 & 8 \\ -2 & 4 & 0 & 6 \\ -2 & 0 & 4 & 6 \\ 0 & 2 & 2 & 4 \end{bmatrix} \qquad (6.6.2)$$

where the pure strategies are in the order listed above. Any strategy $S \in L$ is an equilibrium strategy of Definition 6.2.1 since

$$AS = 2(1, \ 1, \ 1, \ 1).$$

However, S is not in an ES Set since, for example,

$$e_2 \cdot Ae_2 = 4 \quad \text{and} \quad S \cdot Ae_2 = S \cdot (2, \ 4, \ 0, \ 2) = 2.$$

Thus, the initial heuristic that the population evolves to a strategy independent of ownership position is incorrect from the single-species extensive-game perspective.

On the other hand, it is readily apparent that e_2 and e_3 are ESS's for (6.6.2) because the diagonal entries are largest in these two columns. Maynard Smith (1982) called these the *common-sense ESS* and *paradoxical ESS* respectively. As he argues, it is hard to conceive of a biological species evolving to the paradoxical ESS that implies each individual alternates without a fight between being an owner and then an intruder etc. in successive contests.

In fact, by Theorem 6.2.5, there are no other ES Sets. If there were, then it would be of the form L_S for some S with support contained in $\{1, \ 4\}$. Since (6.6.1) is the corresponding 2×2 subgame of (6.6.2), S must be $(1/2, \ 0, \ 0, \ 1/2)$. Although S is an equilibrium strategy, it is not in an ES Set because $S \cdot Ae_2 = 2$ is less than $e_2 \cdot Ae_2 = 4$.

Our intuition now suggests an actual population engaged in an owner-intruder game will evolve to the common-sense ESS where owners are aggressive and intruders are not. This is borne out by the following stability analysis for the haploid single-species dynamic (6.3.1).

The dynamic is identical to that given by the symmetric payoff matrix

$$B = \begin{bmatrix} 0 & 2 & 2 & 4 \\ 2 & 4 & 0 & 2 \\ 2 & 0 & 4 & 2 \\ 4 & 2 & 2 & 0 \end{bmatrix}$$

obtained from (6.6.2) by adding 4 (respectively, subtracting 4) to every entry in column 1 (respectively, column 4). As in Chapter 4 (see Theorem 4.1.1 on natural selection), the mean payoff $S \cdot BS$, defined for $S \in \Delta^4$, is a strict Liapunov function for the dynamic induced through (6.3.1).

In particular, for the pure-strategy dynamic, the only equilibrium points are the four pure strategies together with all points in L. The evolutionary outcome is apparent from the result (Cressman 1992) that $\dot{p}_2 > \dot{p}_3$ if and only if $p_2 > p_3$. Thus, all initial polymorphic populations evolve to e_2 if p_2 is initially larger than p_3; to e_3 if $p_2 < p_3$; and to a point in L if $p_2 = p_3$ initially. That is, any population where owners are initially, on average, more aggressive than intruders will evolve to the common-sense ESS.

(B) The Two-Species Perspective

The single-species extensive-game perspective explicitly assumes each individual must have a strategy for both ownership possibilities (i.e. when owner or intruder). However, at the evolutionarily stable common-sense ESS, an owner will always be an owner. This raises the biological question of how these individuals can pass on to their offspring the strategy to use if they are intruders. Perhaps the model is most realistic in a discrete-generation situation where the offspring of all individuals have an equal chance of being the territorial owner. For example, this may happen for a species that has a limited number of nesting sites that are reoccupied randomly at each generation.

On the other hand, in many populations, it seems just as valid to view owners and intruders as separate species whose offspring have the same ownership position as their parent. In this scenario, we have a bimatrix game as in Section 3.3 (A). We will apply the two-species frequency method of Chapter 3 to the owner-intruder game and compare the result to the intuition developed so far.

Let species I be owners and J be intruders. Both interspecific payoff matrices, B and C, are given by (6.6.1). By Theorem 3.4.2, an ESS (S^*, T^*) must have both S^* and T^* as pure strategies (i.e. either H or D) in Δ^2. Of the four possible combinations, $S^* = H$ and $T^* = D$ as well as $S^* = D$ and $T^* = H$ are the only ESS's. These are precisely the common-sense ESS and paradoxical ESS respectively. That is, static ESS theory has produced the same result in this example from both game-theoretic perspectives.

Comparison of the dynamics is also informative. Let p be the frequency of Hawk owners and q of Hawk intruders. As in (3.2.8), the pure-strategy haploid dynamic is

$$\dot{p} = p(1 - p)(1, -1) \cdot B(q, 1 - q)$$
$$= p(1 - p)(4 - 8q) \tag{6.6.3}$$
$$\dot{q} = q(1 - q)(4 - 8p).$$

The flow on the unit square, shown in Figure 6.6.2, indicates clearly that the two ESS's (i.e. $p = 1$, $q = 0$ and $p = 0$, $q = 1$) are both l.a.s. Moreover, if $p > q$ initially, then the population evolves to the common-sense ESS. [If $p = q$ initially, both species evolve to the mixed strategy $S_0^* = (1/2, 1/2)$ that is an unstable saddle node in Figure 6.6.2.]

REMARK 6.6.1. Although the dynamical outcomes of both evolutionary games (single- and two-species) are identical in biological terms for the owner-intruder game, do not conclude that the actual flows are the same. For example, if the behavior of an owner in the single-species perspective is statistically dependent on this individual's strategy as an intruder, then the evolution of Hawk owners will involve more than p and q as in (6.6.3). [A similar situation

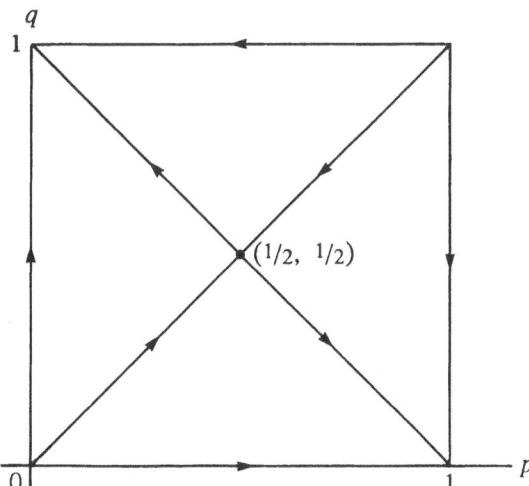

Figure 6.6.2. *The flow diagram for the dynamic (6.6.3)*

occurs for two different mixed-strategy dynamics based on the same payoff matrix where the mean strategy approaches an ESS S^* $\in$ Δ^m in both cases but with different trajectories.] This illustrates again the power of evolutionary game theory to predict evolutionary outcomes without recourse to the particular dynamic.

REMARK 6.6.2. Some readers may be concerned that, in the two-species perspective, all owners at the common-sense ESS have fitness 8 while all intruders have fitness 0. That is, won't the intruders go extinct? This concern goes to the very heart of the two-species frequency-evolution approach where it is assumed both species maintain an equal number of individuals (or at least these two numbers are in fixed ratio). An alternative approach, not pursued in this text, would incorporate density effects to generalize the method of Chapter 5 to two-species models.

REMARK 6.6.3. Evolutionary games based on a single contest having informational asymmetry can always be modelled as a bimatrix game where payoffs refer to interspecific interactions (the perspective of this subsection). On the other hand, single contests involving only environmental information for both contestants (e.g. the model of Section 6.4) can always be modelled by multi-species games having only intraspecific interactions. To see this, define a "species" as all those individuals who find themselves in the same environmental situation. In an example with two possible environments, the two-species model of Chapter 3 then has A and D as the only nonzero payoff matrices. From Definitions 3.4.1 and 2.4.2, it is apparent that (S^*, T^*) is now a two-species ESS if and only if S^* is an ESS for A and T^* is an ESS for D. In particular, the behavior-strategy description of the ES Set for the hawk-dove game with varying resource (Section 6.4) becomes a two-species ESS is this perspective.

REMARK 6.6.4. By combining Remarks 6.6.1 and 6.6.3, the reader would certainly be justified in questioning the necessity of introducing the extensive form for evolutionary games. There are a number of reasons I personally feel the extensive form is important. First, it provides a better intuitive (and visual) understanding of the game, especially when the biological basis for the model rests on a single species. If the dynamical flow is of interest in these cases (and not simply the evolutionary outcome), then the extensive-game dynamic is more appropriate and, at the same time, allows genetic effects such as those considered in Chapter 4 to be included more naturally. Second, if payoffs based on pairwise confrontations result from a series of contests, the extensive form may be unavoidable. The first example in the following section is one such game.

6.7 Multi-Stage Games

Currently, one of the most active research areas in theoretical evolutionary games is the analysis of single-species models where both contestants know if they have been engaged in previous confrontations and, if so, remember what strategies were used in them. There is obvious biological relevance to such models since individuals in a population typically do interact in more than one way — they may either successively or concurrently compete over mates, over nesting sites, over resources etc. In general, the list of possible strategies and/or payoff-matrix entries from one competition may have no relation to those of another.

However, many of the theoretical complications and insights for general multi-stage games are already apparent in games where pairs of individuals are involved in successive competitions that are modelled by the same payoff matrix; each competition becoming a *trial* for the complete game. [Alternatively, we say that the model is an *iterated* or *repeated* game.]

One of the most widely studied models (by classical and biological game theorists alike) is the iterated prisoner's dilemma (Axelrod 1984). It is a theoretical model that analyzes the emergence of cooperative behavior in populations where an individual's best interests in a single trial imply no cooperation should occur. Before developing a multi-trial model of the iterated prisoner's dilemma in the second part of this section, I will first use the extensive form to analyze a simpler two-trial game.

(A) A Two-Trial Game

In this example, opponents in pairwise contests are chosen at random from the population. Each contest involves two trials in which the two contestants choose strategies from the same set of two pure strategies. In the first trial, their choice is made without knowing the opponent's strategy. However, on the second trial, an individual may base its decision on the strategies used in the first trial. An individual's payoff for the complete game is cumulative (i.e. the sum of its payoffs in the two trials).

For mathematical reasons (specifically to introduce only two parameters, a and b, while maintaining sufficient generality to generate interesting phenomena), suppose the 2×2 payoff matrix for each trial is

$$\begin{array}{cc} & \begin{array}{cc} C & D \end{array} \\ \begin{array}{c} C \\ D \end{array} & \begin{bmatrix} 0 & a \\ b & 0 \end{bmatrix}. \end{array} \qquad (6.7.1)$$

The pure strategies are labelled C and D for reasons that will become clear in part B of this section.

The game in extensive form is shown in Figure 6.7.1. For instance, the payoff for the play ending at z_6 is $\begin{vmatrix} 2a \\ 2b \end{vmatrix}$ since player 1 uses C in both trials against D for a total payoff of $a + a$. Also noteworthy is the symmetry indicated at the second trial has player 1 and player 2 in the same situation. That is, at both x_3 and y_4, the player's decision is based on knowing it chose C and the opponent chose D in the first trial.

Since there are two choices at each of the five information sets for either player, a behavior strategy for, say, player 1 can be represented in the form

$$b = (b_1; b_2, b_3, b_4, b_5)$$

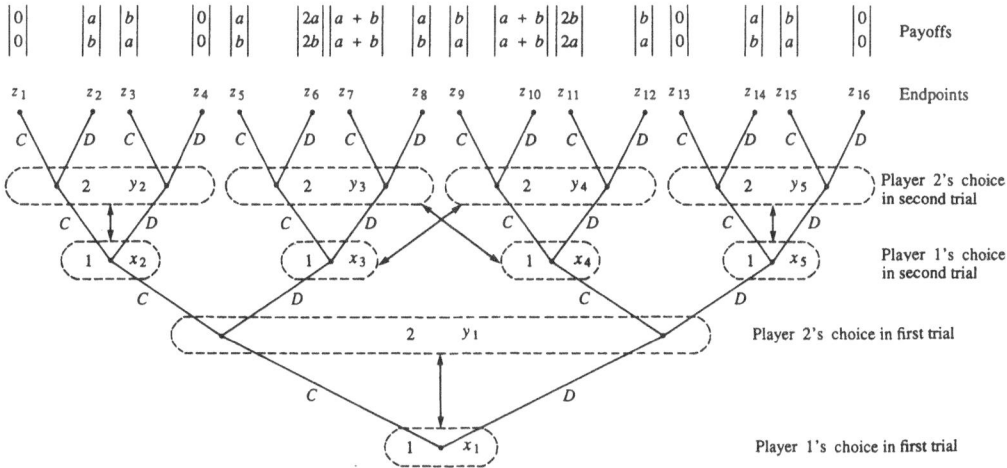

Figure 6.7.1. *The two-trial game in extensive form*

where b_i is the probability of using C at decision point x_i. Probability b_1 is separated from the other four to remind us it is the only one relevant for the first trial.

The extensive form is often condusive to the determination of direct ESS's. The method for this example goes as follows. Consider the four *subgames* Γ_2, Γ_3, Γ_4 and Γ_5 of Figure 6.7.1 that have starting points x_2, x_3, x_4 and x_5 respectively. [The term subgame here refers to an extensive game and should not be confused with earlier usage in this text where it referred to principal submatrices of the normal form.] The subgame Γ_2 (and Γ_5) is a symmetric extensive game corresponding to a single trial with payoff matrix (6.7.1). In this situation, the method asserts a direct ESS b^* must have b_2^* (and b_5^*) as the first component of an ESS of (6.7.1). The game tree in Figure 6.7.1 can then be *reduced* by replacing Γ_2 (and Γ_5) by endpoints at x_2 (and x_5) with payoffs given by the players adopting this ESS strategy in the second trial.

On the other hand, Γ_3 and Γ_4 are not symmetric games but do combine to form a game tree identical to the owner-intruder game of Figure 6.6.1 except for the initial random decision. Nevertheless, b^* must have b_3^* and b_4^* given by the components of a direct ESS for the one-trial symmetric owner-intruder game. Now, Figure 6.7.1 can be reduced to a simple one-trial evolutionary game as in Figure 6.7.2 that yields b_1^* as the first component of its ESS.

To illustrate the method, let us consider the most interesting case for the payoff matrix (6.7.1); namely, a and b are both positive. Here Γ_2 (and Γ_5) has a unique ESS

$$S^* = \left(\frac{a}{a+b} , \frac{b}{a+b} \right) \tag{6.7.2}$$

and this implies $b_2^* = b_5^* = \dfrac{a}{a+b}$. The payoff to both players in these subgames is $ab/(a+b)$ as in Figure 6.7.2. The symmetric game combining Γ_3 and Γ_4 has, with $a > 0$ and $b > 0$, exactly two direct ESS's; namely

(i) Choose C at x_3 and D at x_4.

(ii) Choose D at x_4 and C at x_3.

These ESS's correspond to the common-sense and paradoxical ESS's of Section 6.6. [To prove this, one calculates the 4×4 payoff matrix in normal form for this game in much the same way as (6.6.2) follows from Figure 6.6.1.]

A useful description (and also a way to distinguish these two direct ESS's) is to label them as (i) "repeating" and (ii) "reversing" in that (i) repeats the player's choice from the first trial while (ii) reverses it. In summary, we have shown so far that a direct ESS of the complete game has one of the following two forms.

115

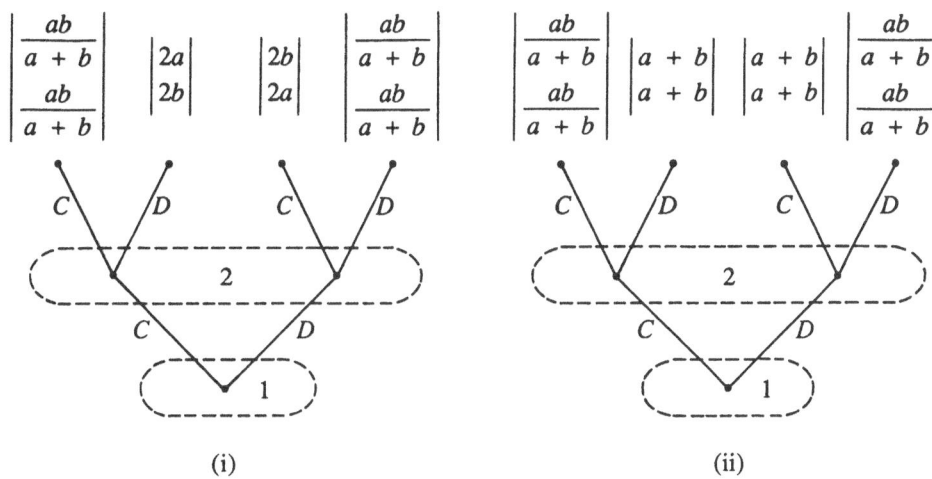

Figure 6.7.2. *The reduced form of a two-trial game*

When a and b are both positive, the two-trial game in extensive form can be reduced to the first trial. Players either use (i) the repeating ESS or (ii) the reversing ESS at the second trial.

(i) $\quad b^* = \left(\quad ; \dfrac{a}{a + b}, \; 1, \; 0, \; \dfrac{a}{a + b} \right)$

(ii) $\quad c^* = \left(\quad ; \dfrac{a}{a + b}, \; 0, \; 1, \; \dfrac{a}{a + b} \right).$

Let us now determine the first components of b^* and c^*. In the reduced game, Figure 6.7.2, the payoffs at x_3 and x_4 are $\begin{vmatrix} 2a \\ 2b \end{vmatrix}$ and $\begin{vmatrix} 2b \\ 2a \end{vmatrix}$ respectively for b^* and are $\begin{vmatrix} a + b \\ a + b \end{vmatrix}$ and $\begin{vmatrix} a + b \\ a + b \end{vmatrix}$ respectively for c^*. The payoff matrices for the reduced games in normal form,

$$\begin{bmatrix} \dfrac{ab}{a + b} & 2a \\ 2b & \dfrac{ab}{a + b} \end{bmatrix} \quad \text{and} \quad \begin{bmatrix} \dfrac{ab}{a + b} & a + b \\ a + b & \dfrac{ab}{a + b} \end{bmatrix}$$

corresponding to b^* and c^* respectively, have the unique ESS's $k(a(2a + b), b(a + 2b))$ and $(1/2, 1/2)$ respectively where $k = 2a^2 + 2ab + 2b^2$. That is, whenever $a > 0$ and $b > 0$, the complete game has exactly two direct ESS's; namely

(i) $\quad b^* = \left(ka(2a + b); \; \dfrac{a}{a + b}, \; 1, \; 0, \; \dfrac{a}{a + b} \right)$ \hfill (6.7.3)

(ii) $\quad c^* = \left(1/2; \; \dfrac{a}{a + b}, \; 0, \; 1, \; \dfrac{a}{a + b} \right).$ \hfill (6.7.4)

Both these direct ESS's can be given a heuristic interpretation based on rational decision processes. Each player would prefer different strategies be used on the first trial to avoid the 0 payoff. However, if the same strategies are used, neither player knows if its opponent will switch strategies so that both should adopt a strategy in the second trial that offers no incentive for the opponent to alter its choice. That is, the second and fifth components of the direct ESS should be the first component of (6.7.2).

On the other hand, if opposite strategies are used in the first trial, a player can either be "sharing" or "selfish". Let us assume the positive payoffs satisfy $a > b$ from now on. Then both players would like to receive payoff a at each trial but also realize the opponent feels the same. A sharing individual would settle for the next best outcome, that has a at one trial and b at the other, not caring in what order they occur. This is c^* that has $c_1^* = 1/2$. However, a selfish individual tries to obtain a at both trials by choosing C more frequently at the first trial. In fact, it is true that b^* chooses C more frequently than even the single-trial ESS of (6.7.2) since $b_1^* > \dfrac{a}{a + b} > 1/2$. Then, if the selfish individual wins (i.e. obtains payoff a), it tries to repeat its success at the second trial by again choosing C.

It is interesting to note that monomorphic populations using b^* expect a payoff of $a + b$ in those plays that have different strategies at the first trial as do individuals in a c^* monomorphic population. That is, sharing and selfish populations do equally well in this part of the game tree. However, overall the sharing population does better since the 0 payoff is reached only half the time for it as opposed to a proportion $(b_1^*)^2 + (1 - b_1^*)^2 > 1/2$ of the time for a selfish population. This suggests sharing populations may be more likely to evolve. The dynamic implications will be discussed further after the normal form is developed.

For the less-interesting cases of (6.7.1) (i.e. either a or b is nonpositive), there are no direct ESS's. However, there are ES Sets in the normal form of this extensive game (these correspond to what could be called direct ES Sets). Technically, there are 32 (i.e. 2^5) pure strategies for the extensive game; each of which specifies a firm action, either C or D, at a player's five information sets. However, some pure strategies are indistinguishable. For example, if player 1 uses C in the first trial, it is meaningless to specify actions at x_4 and x_5 in Figure 6.7.1. By this means the number of pure strategies is reduced to 8 (i.e. 2^3).

A pure strategy will be denoted by $[\alpha; \beta, \gamma]$ where α, β and γ are either C or D; α specifies the action in the first trial; either β or γ specifies the action in the second trail depending on the opponent's choice C or D respectively in the first trial. With the pure strategies given in the order $[C; C, C]$, $[C; C, D]$, $[C; D, C]$, $[C; D, D]$, $[D; C, C]$, $[D; D, C]$, $[D; C, D]$ and $[D; D, D]$, the 8×8 payoff matrix is

$$A = \begin{bmatrix} 0 & 0 & a & a & a & 2a & a & 2a \\ 0 & 0 & a & a & a+b & a & a+b & a \\ b & b & 0 & 0 & a & 2a & a & 2a \\ b & b & 0 & 0 & a+b & a & a+b & a \\ b & a+b & b & a+b & 0 & 0 & a & a \\ 2b & b & 2b & b & 0 & 0 & a & a \\ b & a+b & b & a+b & b & b & 0 & 0 \\ 2b & b & 2b & b & b & b & 0 & 0 \end{bmatrix}. \tag{6.7.5}$$

Cannings and Whittaker (1991) classified the ESS structure of A. If either a or b is nonpositive, A has no ESS. However, there are ES Sets. When $a < 0$ and $b < 0$, there are exactly four; namely,

$$\{(p, \ 1-p, \ 0, \ 0, \ \cdots, 0) \in \Delta^8 \mid 0 \le p \le 1\} \tag{6.7.6a}$$

$$\{(0, \ 0, \ p, \ 1-p, \ 0, \ 0, \ 0, \ 0) \mid 0 \le p \le 1\} \tag{6.7.6b}$$

$$\{(0, \ 0, \ 0, \ 0, \ p, \ 1-p, \ 0, \ 0) \mid 0 \le p \le 1\} \tag{6.7.6c}$$

$$\{0, \ \cdots, 0, \ p, \ 1-p) \mid 0 \le p \le 1\}. \tag{6.7.6d}$$

Thus any population with mean strategy in an ES Set has all contestants choosing the same strategy (i.e. either C or D which correspond to the ESS's of (6.7.1)) in a given trial of the complete game. These ES Sets do not correspond to direct ESS's since, for example, (6.7.6a) is equivalent to the set of behavior strategies that have $b_1 = b_2 = 1$. When $a < 0$ and $b > 0$, (6.7.6d) is the only ES Set (here individuals always choose D — the only ESS of (6.7.1)). These results are all consistent with the intuition behind a one-trial game.

Let us return to the most interesting case (i.e. $a > 0$ and $b > 0$). Cannings and Whittaker (1991) showed A now has exactly two ESS's (their method also shows A has no other ES Sets). These are

$$p^* = \frac{k}{a+b} (a^2(2a+b), \ 0, \ ab(2a+b), \ 0, \ 0, \ ab(a+2b), \ 0, \ b^2(a+2b)) \tag{6.7.7}$$

$$q^* = \frac{1}{2(a+b)} (0, \ a, \ 0, \ b, \ a, \ 0, \ b, \ 0) \tag{6.7.8}$$

where we recall $k = 2a^2 + 2ab + 2b^2$. [Note that I have adopted their notation instead of using S^* and T^* to denote ESS's.]

In my opinion, their approach is backwards since their p^* and q^* appear as if by magic before the behavior-strategy description of these ESS's is briefly mentioned. It is easy to see p^* corresponds to b^* (and q^* to c^*) by verifying the following five required equations are true.

$$b_1^* = p_1^* + p_2^* + p_3^* + p_4^*; \quad b_1^* b_2^* = p_1^* + p_2^*; \quad b_1^* b_3^* = p_1^* + p_3^*;$$

$$(1 - b_1^*)b_4^* = p_5^* + p_7^*; \quad (1 - b_1^*)b_5^* = p_5^* + p_6^*.$$

Conceptually, it is better in this example to first find b^* and then solve for p^* through these five equations.

On the other hand, the normal form is condusive to analyzing the dynamic properties of the ESS structure. For instance, if $a < 0$ and $b > 0$, the pure-strategy dynamic on Δ^8

$$\dot{p}_i = p_i(e_i - p) \cdot Ap$$

approaches an equilibrium point in the ES Set (6.7.5d) for any initial polymorphism. If $ab > 0$, the global evolution is not completely understood even for the pure-strategy dynamic. However, we can say that there is a two-dimensional linear submanifold L of Δ^8, given by

$$\{S_0^* + \lambda_1(1, -1, -1, 1, 0, 0, 0, 0) + \lambda_2(0, 0, 0, 0, 1, -1, -1, 1) \in \Delta^8 \mid \lambda_1, \lambda_2 \in R\}$$

where $S_0^* = (1/(a+b)^3)(a^3, a^2b, a^2b, ab^2, a^2b, ab^2, ab^2, b^3)$, that consists entirely of equilibrium strategies. If $a > 0$ and $b > 0$, L appears to attract a six-dimensional submanifold of Δ^8 that separates Δ^8 into two regions, R_1 and R_2, that computer simulations (Cannings and Whittaker 1991) suggest are the respective basins of attraction for p^* and q^*. It also appears that R_2 has the larger (seven-dimensional) volume whenever $a \neq b$. In this sense sharing populations are more likely to evolve than selfish ones.

REMARK 6.7.1. It is possible to develop this two-trial game so that our eight pure strategies come directly from the extensive form without necessitating a reduction from the 32 pure strategies of Figure 6.7.1. This is accomplished by combining, for player 1, x_2 and x_4 in the same information set as well as x_3 and x_5. This models the situation where a player's decision at the second trial depends only on the opponent's choice at the first trial. The three information sets for a player then provide the eight pure strategies considered above. However, player 1 cannot remember its first decision and so the extensive game in this form does not exhibit perfect recall. I have avoided the resultant technical complication by adhering to Figure 6.7.1 instead.

(B) The Iterated Prisoner's Dilemma Game

In each trial of this iterated game, both opponents choose whether to cooperate with each other (C) or defect (D). The payoff matrix in each trial is

$$\begin{array}{cc} & \begin{array}{cc} C & D \end{array} \\ \begin{array}{c} C \\ D \end{array} & \begin{bmatrix} R & S \\ T & P \end{bmatrix} \end{array}. \qquad (6.7.9)$$

where R is the "reward" for mutual cooperation; P the "punishment" for mutual defection; T the "temptation" to defect; and S the "sucker's" payoff. In the prisoner's dilemma, these payoffs are ranked by the order

$$T > R > P > S. \qquad (6.7.10)$$

[Most game theory texts (Luce and Raiffa 1957; Owen 1982; Mesterton-Gibbons 1992) elaborate on the relevance of this order for many other confrontational situations besides that of two prisoners suspected of a crime for which the game is named.]

It is clear that biological species where individual fitness depends on a single random encounter of prisoner's dilemma will evolve to the only ESS of (6.7.9). That is, all individuals will eventually be defectors who then have fitness P. It is also clear that everyone would be better off (i.e. have fitness R) if they all were cooperators. Therein lies the dilemma. Since it is better to defect no matter what your opponent chooses (i.e. $T > R$ and $P > S$), it is difficult to see how a cooperative population could either evolve or maintain itself without appealing to some group selection argument.

Perhaps individual selection can lead to cooperation if the game is repeated. However, rational players who base their choice on strategies used in previous trials will always defect if there are a fixed number of trials known to both contestants and payoffs are cumulative (Luce and Raiffa 1957). This is true of non-rational populations as well. For example, the two-trial prisoner's dilemma can be analyzed by the method in Section 6.7 (A) with (6.7.9) in place of (6.7.1). The only ES Set is (6.7.6d) corresponding to populations where everyone defects. Moreover, this ES Set is a global attractor under the pure-strategy dynamic for any initial polymorphism.

On the other hand, it has been shown that more cooperative strategies may emerge if the number of trials is a probability distribution (Mesterton-Gibbons 1992) or, equivalently, future payoffs are discounted using a certain rate of inflation. This approach assumes that the number N of trials is large (say, $N > 30$) or infinite and that contestants remember all previous mutual encounters. Either of these assumptions create mathematical and/or conceptual difficulties. Mathematically, the list of pure strategies becomes astronomical (greater than $2^{(2^{30})}$). Many studies have then focussed on the stability of a cooperative strategy S_0 against a restricted set of pure strategies (e.g. the evolutionary stability of S_0 under monomorphic invasion). The most widely analyzed S_0 of this type is TFT (Tit-for-Tat) which cooperates on the first trial and repeats the opponent's choice on each successive trial. Conceptually, the non-rationality assumption of evolutionary game theory would appear to be violated by strategies based on outcomes of 30 previous trials.

A reasonable compromise, in keeping with the spirit of this text, is to investigate evolutionary stability under invasion by arbitrary polymorphisms assuming strategies are based on a limited amount of information from a large number of previous encounters. Specifically, we will assume for the rest of this section that, after the first trial, strategies depend solely on the opponent's last choice and that payoffs for the complete game are given as the (limiting) average payoff over a large number of trials. *TFT* is then a legitimate pure strategy (it is the second on the list in Section 6.7 (A)) while a strategy that alternates between C and D, no matter what the opponent does, is not. [See Borštnik et al. (1990) for extensions of this example that include such alternating strategies.]

The extensive form of the game analogous to Figure 6.7.1 is not very useful here because the game does not exhibit perfect recall (e.g. an individual's choice at the third trial cannot depend on what it chose at the first two) and the payoffs are not cumulative. In normal form, the 8×8 payoff matrix A (using the same order for pure strategies as part A) is

$$
\begin{bmatrix}
R & R & S & S & R & S & R & S \\
R & R & U & P & R & U & 1/2(S+T) & P \\
T & U & 1/2(R+P) & S & T & S & U & S \\
T & P & T & P & T & T & P & P \\
R & R & S & S & R & S & R & S \\
T & U & T & S & T & 1/2(R+P) & U & S \\
R & 1/2(S+T) & U & P & R & U & P & P \\
T & P & T & P & T & T & P & P
\end{bmatrix}
\qquad (6.7.11)
$$

where $U = 1/4(R + S + T + P)$. Let us check entries A_{26} and A_{62} (the reader should check some of the other 62 entries). When $[C; C, D]$ plays $[D; D, C]$, the first trial is CD (i.e. C is $[C; C, D]$'s choice); DD is the second trial; DC the third; CC the fourth; CD the fifth; and the cycle continues from there. Thus, on average, both individuals receive payoff $1/4(R + S + T + P)$.

From now on, assume that mutual cooperation has a higher reward then alternating trials of CD and DC. That is, assume

$$
R > \frac{S + T}{2}. \qquad (6.7.12)
$$

The following result will be disappointing for those who hoped static ESS theory would "solve" this example.

THEOREM 6.7.2. The payoff matrix (6.7.11) has no ES Set.

PROOF. From (6.7.11), it is clear that strategies e_1 and e_5 are indistinguishable in normal form as are e_4 and e_8. Thus the fourth and fifth rows and columns can be deleted from (6.7.11) to form a reduced 6×6 payoff matrix. Suppose L is an ES Set of this reduced matrix. The proof is divided into parts.

(i) $\{e_1, e_2\} \cap L = \emptyset$. If $S_0 \in L$, then $2 \notin$ supp S_0.

    Proof. $e_1 \notin L$ since $e_8 \cdot Ae_1 > e_1 \cdot Ae_1$. If $e_2 \in L$, then $e_1 \in L$ because $e_1 \cdot Ae_2 = e_2 \cdot Ae_2$ and $e_2 \cdot Ae_1 = e_1 \cdot Ae_1$.

(ii) $\{e_7, e_8\} \cap L = \emptyset$. If $S_0 \in L$, then $8 \notin \text{supp} S_0$.

Proof. Similar to (i).

(iii) If $S_0 \in L$, then $7 \in \text{supp} S_0$.

Proof. If $1 \in \text{supp} S_0$, then $e_1 \cdot AS_0 = S_0 \cdot AS_0$. Thus $S_0 \cdot Ae_1 > e_1 \cdot Ae_1$. From A_{11}, A_{31}, A_{61} and A_{71}, we see either 3 or 6 belongs to $\text{supp} S_0$. Similarly, if $3 \in \text{supp} S_0$, either 6 or 7 belongs to $\text{supp} S_0$. Also, if $6 \in \text{supp} S_0$, then $7 \in \text{supp} S_0$.

(iv) If $S_0 \in L$, $7 \notin \text{supp} S_0$.

Proof. If $7 \in \text{supp} S$, $e_8 \cdot AS_0 \leq S_0 \cdot AS_0 = e_7 \cdot AS_0$. Therefore, $S_0 = e_7$. [If this were not the case, then $e_8 \cdot AS_0 = T(s_1 + s_3 + s_6) + Ps_7$ is strictly greater than $e_7 \cdot AS_0 = Rs_1 + 1/4(R + S + T + P)(s_3 + s_6) + Ps_7$.] This contradicts (ii).

By parts (iii) and (iv), the proof is complete.

∎

There are, however, exactly two pure equilibrium strategies (Definition 6.2.1.) of the reduced 6×6 payoff matrix; namely, e_2 and e_8. These are the *TFT* and defect-no-matter-what strategies respectively. Their respective dynamic stability properties will go a long way to understanding the evolution of cooperation in this model. Strategy e_8 is not stable. In fact, a population where everyone always defects can be invaded successfully by a monomorphic *TFT* subpopulation. Eventually, all individuals will play *TFT* at which time all trials will consist of mutual cooperation.

This does not mean *TFT* populations are globally stable. Indeed, if $P > 1/2(S + T)$, then e_7 (the strategy that uses D at the first trial and *TFT* thereafter) cannot be invaded by *TFT*. Nevertheless, if at least half the population are initially *TFT* individuals, then this proportion will never decrease under the haploid dynamic (Borštnik et al. 1990). Their proof for the pure-strategy dynamic, $\dot{p}_i = p_i(e_i - p) \cdot Ap$, shows

$$\dot{p}_2 \geq (2p_2 - 1) \sum_{j \neq 2} p_2 p_j (R - A_{j2}) \qquad (6.7.13)$$

where p_2 is the frequency of *TFT* individuals. [The result remains valid for the mixed-strategy dynamic as well.] That is, p_2 is a strict local Liapunov function.

Moreover, if $p_2 > 1/2$ initially, it is clear from (6.7.11) and (6.7.13) that all individuals will eventually use *TFT* or cooperate-no-matter-what (recall that e_5 was deleted) and so all trials are mutually cooperative. [It should also be mentioned that e_2 is an ESS if e_1 is also deleted.] In the language of dynamical systems, the line segment in Δ^6 joining $1/2(e_1 + e_2)$ to e_2 is a set of neutrally stable equilibria that attracts any initial state with $p_2 > 1/2$. Heuristically, a population will evolve to mutual cooperation and remain there if the population as a whole guards against complacency by maintaining a sufficient number of *TFT* individuals who will punish any mutant defectors.

REMARK 6.7.3. I consider the iterated prisoner's dilemma game a perfect example on which to conclude this text. It illustrates both the power and limitations of ESS theory. My own bias is that dynamical considerations must form an integral part of theoretical (evolutionary) games. Even though, in many models, the solution concept from static ESS theory determines the evolutionary outcome, there are many important biological models (such as those in this section) where dynamical properties are essential to compare different realistic solutions. This bias is one that is becoming more widespread among classical and evolutionary game theorists alike.

REFERENCES

Akin, E. (1979). *The Geometry of Population Genetics*. Lecture Notes in Biomathematics **31**. Springer-Verlag: New York, Heidelberg, Berlin.

Akin, E. (1982). Exponential families and game dynamics. *Can. J. Math.* **34**, 374-405.

Akin, E. (1990). The differential geometry of population genetics and evolutionary games. In *Mathematical and Statistical Developments of Evolutionary Theory* (S. Lessard, Ed.). Kluwer Academic: Dordrecht, 1-93.

Arrow, K.J. and Hahn, F.H. (1971). *General Competitive Analysis*. Holden-Day: San Francisco.

Axelrod, R. (1984). *The Evolution of Cooperation*. Basic: New York.

Bodmer, W.F. and Felsenstein, J. (1967). Linkage and selection: theoretical analysis of the deterministic two locus random mating model. *Genetics* **57**, 237-265.

Bomze, I.M. and Pötscher, B.M. (1989). *Game Theoretical Foundations of Evolutionary Stability*. Lecture Notes in Economics and Mathematical Systems **324**. Springer-Verlag: New York, Heidelberg, Berlin.

Borštnik, B., Pumpernik, D., Hofacker, I.L., and Hofacker, G.L. (1990). An ESS-analysis for ensembles of prisoner's dilemma strategies. *J. Theor. Biol.* **142**, 189-200.

Cannings, C. and Whittaker, J.C. (1991). A two-trial two-strategy conflict. *J. Theor. Biol.* **149**, 281-286.

Carr, J. (1981). *Applications of Centre Manifold Theory*. Springer-Verlag: New York, Heidelberg, Berlin.

Christiansen, F.B. (1991). On conditions for evolutionary stability for a continuously varying character. *Am. Nat.* **138**, 37-50.

Cockerham, C.C., Burrows, P.M., Young, S.S., and Prout, T. (1972). Frequency-dependent selection in randomly mating populations. *Am. Nat.* **106**, 493-515.

Cressman, R. (1988a). Frequency-dependent viability selection (a single-locus, multi-phenotype model). *J. Theor. Biol.* **130**, 147-165.

Cressman, R. (1988b). Complex dynamical behaviour of frequency-dependent viability selection: an example. *J. Theor. Biol.* **130**, 167-173.

Cressman, R. (1988c). Frequency- and density-dependent selection: the two-phenotype model. *Theor. Pop. Biol.* **34**, 378-398.

Cressman, R. (1990). Strong stability and density-dependent evolutionarily stable strategies. *J. Theor. Biol.* **145**, 319-330.

Cressman, R. (1992). Evolutionarily stable sets in symmetric extensive two-person games. *Math. Biosci.* (in press).

Cressman, R. and Dash, A.T. (1987). Density dependence and evolutionary stable strategies. *J. Theor. Biol.* **126**, 393-406.

Cressman, R. and Dash, A.T. (1991). Strong stability and evolutionarily stable strategies with two types of players. *J. Math. Biol.* **30**, 89-99.

Cressman, R., Dash, A.T., and Akin, E. (1986). Evolutionary games and two species population dynamics. *J. Math. Biol.* **23**, 221-230.

Cressman, R. and Hines, W.G.S. (1984). Evolutionarily stable strategies of diploid populations with semi-dominant inheritance patterns. *J. Appl. Prob.* **21**, 1-9.

Dawkins, R. (1976). *The Selfish Gene.* Oxford University Press.

Edgeworth, F.Y. (1881). Mathematical Psychics: An Essay on the Applications of Mathematics to the Moral Sciences. Paul Keegan: London.

Eshel, I. (1982). Evolutionarily stable strategies and viability selection in Mendelian populations. *Theor. Pop. Biol.* **22**, 204-217.

Fisher, R.A. (1930). *The Genetical Theory of Natural Selection.* Clarendon Press: Oxford.

Gayley, T.W. and Michod, R.E. (1990). The modification of genetic constraints on frequency-dependent selection. *Am. Nat.* **136**, 406-427.

Ginzburg, L.R. (1977). The equilibrium and stability for n alleles under the density-dependent selection. *J. Theor. Biol.* **68**, 545-550.

Guckenheimer, J. and Holmes, P.J. (1983). *Nonlinear Oscillations, Dynamical Systems, and Bifurcations of Vector Fields.* Springer-Verlag: New York, Heidelberg, Berlin.

Haigh, J. (1975). Game theory and evolution. *Adv. Appl. Prob.* **7**, 8-11.

Hicks, J.R. (1939). *Value and Capital.* Oxford University Press.

Hammerstein, P. (1990). Evolutionary stability and the streetcar theory of evolution. Presented at the *Animal Conflict Workshop,* Bonn.

Hines, W.G.S. (1980). Strategy stability in complex populations. *J. Appl. Prob.* **17**, 600-610.

Hines, W.G.S. (1987). Evolutionary stable strategies: a review of basic theory. *Theor. Pop. Biol.* **31**, 195-272.

Hines, W.G.S. (1991). ESS modelling of diploid populations I & II. University of Guelph Statistical Series 1991-243/244.

Hines, W.G.S. and Anfossi, D. (1990). A discussion of evolutionarily stable strategies. In *Mathematical and Statistical Developments of Evolutionary Theory* (S. Lessard, Ed.) Kluwer Academic: Dordrecht, 229-267.

Hines, W.G.S. and Bishop, D.T. (1984). Can and will a sexual diploid population attain an evolutionary stable strategy? *J. Theor. Biol.* **111**, 667-686.

Hofbauer, J. and Sigmund, K. (1988). *The Theory of Evolution and Dynamical Systems.* Cambridge University Press.

Kingman, J.F.C. (1961). A mathematical problem in population genetics. *Proc. Camb. Phil. Soc.* **57**, 574-582.

Lessard, S. (1984). Evolutionary dynamics in frequency-dependent two-phenotype models. *Theor. Pop. Biol.* **25**, 210-234.

Lloyd, D.G. (1977). Genetic and phenotypic models of natural selection. *J. Theor. Biol.* **69**, 543-560.

Luce, R.D. and Raiffa, H. (1957). *Games and Decisions.* Wiley: New York.

Maynard Smith, J. (1974). The theory of games and the evolution of animal conflicts. *J. Theor. Biol.* **47**, 209-221.

Maynard Smith, J. (1981). Will a sexual population evolve to an ESS? *Am. Nat.* **117**, 1015-1018.

Maynard Smith, J. (1982). *Evolution and the Theory of Games.* Cambridge University Press.

Maynard Smith, J. (1988). Can a mixed strategy be stable in a finite population? *J. Theor. Biol.* **130**, 247-251.

Maynard Smith, J. and Price, G.R. (1973). The logic of animal conflict. *Nature* **246**, 15-18.

Mesterton-Gibbons, M. (1992). *An Introduction to Game-Theoretic Modelling.* Addison-Wesley: Redwood City.

Michod, R.E. (1984). Genetic constraints on adaptation, with special reference to social behavior. In *The New Ecology: Novel Approaches to Interactive Systems.* Wiley: New York, 253-276.

Motro, U. (1991). Avoiding inbreeding and sibling competition: the evolution of sexual dimorphism for dispersal. *Am. Nat.* **137**, 108-115.

Owen, G. (1982). *Game Theory.* 2nd edition. Academic Press: New York.

Petersen, C.W. (1991). Sex allocation in hermaphroditic sea basses. *Am. Nat.* **138**, 650-667.

Pielou, E.C. (1977). *Mathematical Ecology.* Wiley: New York.

Riechert, S. and Hammerstein, P. (1983). Game theory in the ecological context. *Ann. Rev. Ecol. Syst.* **14**, 377-409.

Roughgarden, J. (1979). *Theory of Population Genetics and Evolutionary Ecology: An Introduction.* Macmillan: New York.

Rowe, G.W., Harvey, I.F., and Hubbard, S.F. (1985). The essential properties of evolutionary stability. *J. Theor. Biol.* **115**, 269-285.

Samuelson, P.A. (1941). The stability of equilibrium: comparative statics and dynamics. *Econometrica* **9**, 97-120.

Samuelson, P.A. (1944). The relation between Hicksian stability and true dynamic stability. *Econometrica,* **12**, 256-257.

Schuster, P., Sigmund, K., Hofbauer, J., Gottlieb, R., and Merz, P. (1981). Self-regulation of behaviour in animal societies III. *Biol. Cybern.* **40**, 17-25.

Selten, R. (1980). A note on evolutionarily stable strategies in asymmetrical animal conflicts. *J. Theor. Biol.* **84**, 93-101.

Selten, R. (1983). Evolutionary stability in extensive two-person games. *Math. Social Sciences* **5**, 269-363.

Shubik, M. (1959). Edgeworth market games. In *Contributions to the Theory of Games, Volume 4* (R.D. Luce and A.W. Tucker, Eds.). Princeton University Press: Princeton.

Shubik, M. (1984). *A Game-Theoretic Approach to Political Economy.* Volume **2**. M.I.T. Press: Cambridge, Mass.

Slatkin, M. (1979). The evolutionary response to frequency- and density-dependent interactions. *Am. Nat.* **114**, 384-398.

Taylor, P.D. (1979). Evolutionarily stable strategies with two types of players. *J. Appl. Prob.* **16**, 76-83.

Taylor, P.D. and Jonker, L. (1978). Evolutionarily stable strategies and game dynamics. *Math. Biosci.* **40**, 145-156.

Thomas, B. (1985a). On evolutionarily stable sets. *J. Math. Biol.* **22**, 105-115.

Thomas, B. (1985b). Evolutionarily stable sets in mixed-strategist models. *Theor. Pop. Biol.* **28**, 332-341.

Thomas, B. (1985c). Genetical ESS-models. I. Concepts and basic model. *Theor. Pop. Biol.* **28**, 18-32.

Thomas, L.C. (1984). *Games, Theory and Applications.* Wiley: New York.

Vickers, G.T. and Cannings, C. (1988). Patterns of ESS's I. *J. Theor. Biol.* **132**, 387-408.

Wiggins, S. (1990). *Introduction to Applied Nonlinear Dynamical Systems and Chaos.* Springer-Verlag: New York, Heidelberg, Berlin.

Zeeman, E.C. (1979). Population dynamics from game theory. *Proc. Int. Conf. Global Theory of Dynamical Systems.* Northwestern: Evanston, 471-497.

Zeeman, E.C. (1981). Dynamics of the evolution of animal conflicts. *J. Theor. Biol.* **89**, 249-270.

INDEX

General Remarks

Lecture Notes are printed by photo-offset from the master-copy delivered in camera-ready form by the authors of monographs, resp. editors of proceedings volumes. For this purpose Springer-Verlag provides technical instructions for the preparation of manuscripts. Volume editors are requested to distribute these to all contributing authors of proceedings volumes. Some homogeneity in the presentation of the contributions in a multi-author volume is desirable.

Careful preparation of manuscripts will help keep production time short and ensure a satisfactory appearance of the finished book. The actual production of a Lecture Notes volume normally takes approximately 8 weeks.

Authors of monographs receive 50 free copies of their book. Editors of proceedings volumes similarly receive 50 copies of the book and are responsible for redistributing these to authors etc. at their discretion. No reprints of individual contributions can be supplied. No royalty is paid on Lecture Notes volumes.

Volume authors and editors are entitled to purchase further copies of their book for their personal use at a discount of 33.3 %, other Springer mathematics books at a discount of 20 % directly from Springer-Verlag. Authors contributing to proceedings volumes may purchase the volume in which their article appears at a discount of 20 %.

Commitment to publish is made by letter of intent rather than by signing a formal contract. Springer-Verlag secures the copyright for each volume.

Addresses:

Professor Simon A. Levin, Princeton University
Department of Ecology and
Evolutionary Biology
Eno Hall, Princeton
New Jersey 08544-1003, USA

Springer-Verlag, Mathematics Editorial
Tiergartenstr. 17
W-6900 Heidelberg
Federal Republic of Germany
Tel.: *49 (6221) 487-410